DATA ANALYSIS IN MOLECULAR BIOLOGY AND EVOLUTION

DATA ANALYSIS IN MOLECULAR BIOLOGY AND EVOLUTION

by

Xuhua Xia
University of Hong Kong

SPRINGER SCIENCE+BUSINESS MEDIA, LLC

Library of Congress Cataloging-in-Publication Data

Xia, Xuhua, 1959-
Data analysis in molecular biology and evolution / by Xuhua Xia.
p. cm.
Includes bibliographical references (p.).
ISBN 978-1-4757-8435-0 ISBN 978-0-306-46893-3 (eBook)
DOI 10.1007/978-0-306-46893-3
1. Molecular biology--Data processing. 2. Evolution (Biology)--Data processing. I. Title.

QH506 .X53 2000
572.8'0285--dc21

00-022032

Originally published by Kluwer Academic Publishers in 2000
Softcover reprint of the hardcover 1st edition 2000

Printed on acid-free paper.

Contents

Acknowledgements

It would have been much easier for me to write this ACKNOWLEDGEMENT if I were a well established scientist of international fame. I could then write in a pastoral manner about sweet recollections of the past, starting with a certain scientist, also internationally famous of course, who came to visit my lab and suggested that I should write such a book. Knowing that the whole world was watching and waiting, I had set aside all the other very important works and devoted most of my time to the writing of this path-blazing masterpiece. Every draft chapter was snatched away by a whole wolf pack of world authorities who would then excitedly share it with their colleagues, postdoctoral fellows and students. Comments and suggestions were then poured in, ultimately leading to this polished gem now resting in your hands. The ACKNOWLEDGMENT could then be optionally concluded with a confident "Please read the book."

But I am neither well established nor internationally famous, and writing the book, as well as the computer program called DAMBE, is mostly my own idea. Few people would be watching and waiting when I wrote the book, and you are likely one of the first few people who accidentally stumbled onto the book, several years after its publication. So my acknowledgement, first of all, goes to you. Thanks for reading the book.

It would be very ungrateful of me if I failed to acknowledge the fact that the book and the program would not have come to their current states without the help and encouragement from many friends and colleagues. However, it is quite awkward for a junior scientist like me to acknowledge contributions from well established senior scientists because it may well be construed as an attempt to boost my low credit rating. So I will write quietly, with no fanfare, that there is indeed a highly respected scientist (also a friend and mentor), who reviewed the first draft and had encouraged me to write the book. In particular, I have benefited greatly from reading his book on molecular evolution, which he gave me as a gift. It has been my dream to be able to give him, as a gift, a book of my own.

There is also another friend and colleague, visiting Hong Kong from Uppsala, who volunteered to read every chapter that I had finished writing. Martin Lascoux, who is at roughly the same credit rating as I am, has been extremely helpful in many ways. Thank you, Martin, for your time and for the many equations you wrote on the back of the manuscript.

My thanks should also go to the many colleagues who used DAMBE and offered me feedback. They are Thomas A. Artiss, A. R. Bensen, James W. Borrone, Carlos Bustamante, Fernando Gonzalez Candelas, T. Y. Chiang, Geoff Clarke, Rich Cronn, Katherine Dunn, Vladimir Dvornik, Ananias A. Escalante, Roger Francis, Thomas Guebitz, Gunther Franz Manni, Gregor Hagedorn, Healy Hamilton, K. Y. Hu, Peter Hughes, Bob Krebs, Konstantin Krutovskii, Richard McCaman, Horacio Naveira, Enrico Negrisolo, Johan Nylander, Jes Soee Pedersen, Stuart Piertney, Henryk Rozycki, Marco Salemi, David Schultz, Gaofeng Shang, Mike Smith, Ulf Sorhannus, Chen Su, Andrea Taylor, Fredj Tekaia, Rodrigo Vidal, Cathy Walton, John Wetherall, Jonathan F Wendel, Tony Wilson, Avshalom Zoossmann, Dmitrij Zubakov. In particular, I wish to thank Tony Wilson for his being the first person to test my program, Gregor Hagedorn for sending me a five-page report on how the program could be improved, Mike Smith for his comments on the program and for his encouragement on writing this book, and Chen Su who is the first Chinese colleague who sent me encouragement on DAMBE development. Please keep in touch.

My program DAMBE has incorporated codes from various other programs: PHYLIP, PAML, ClustalW and a program written by Andrei Zharkikh. I am grateful to the programmers who have made their programs freely available, and I think that the best way for me to show my appreciation for their effort is to make my own program freely available to the scientific community.

Just like all the caring parents who nervously send off their children to brave the real world, I am now, with great anxiety, dispatching my book and the program to explore the unpredictable academic terrain. I am consciously aware that they may subsequently get lost in the wilderness and become homeless. It is exactly for this reason that I wish to thank you again for holding the book with caring hands. May the book and the program be useful to you!

Preface

People learn by observing things around them. When the telescope and the microscope were invented, people aimed them at different objects, large and small, and discovered a new world that had been hidden from them. Interesting patterns gradually take shape and theories gradually come into being, through innovative ways of looking at things.

A computer program for data analysis is analogous to a telescope or a microscope. We use the program to look at the data set, to reveal the patterns that have been hidden from us, and to derive new insights that would otherwise be beyond our imagination. The computer program (DAMBE) that I am promoting in this book is for data analysis in molecular biology, ecology, and evolution, and I hope that it will help you see interesting patterns that have been hidden from you.

The last decade has witnessed an explosive growth of molecular data which, according to bioinformaticians, will be the most important resources in the next century. However, after travelling along the so-called information superhighway for some time, most of us have come to realize that information is not equivalent to knowledge. Indeed, an overwhelming amount of undigested information may not only dazzle our eyes, but also confuse our mind. It is for this reason that many computer programs have been developed in the last decade to facilitate our effort to extract valuable knowledge from the bewildering jungle of information. DAMBE is one of such programs, and this book will take advantage of the powerful analytical features in DAMBE to illustrate innovative ways of treasure hunting in the field of molecular evolution and computational molecular biology.

The book is structured in five parts. Chapter 1 provides a brief introduction to DAMBE, a user-friendly computer program for molecular data analysis. Chapters 2-5 cover routine techniques for retrieving, manipulating, converting, organizing, and aligning molecular sequence data. Chapters 6-11 introduce the concept of a substitution model which typically has two categories of parameters called frequency parameters and rate ratio parameter. The emphasis is on factors that affect the frequency parameters and lead to nucleotide, codon and amino acid usage bias. Recent studies on the effect of maximizing transcriptional and translational efficiencies on codon usage bias were described in detail in an effort to guide the reader to problems that remain unsolved. Chapters 12-16 cover fundamentals of comparative sequence analysis, with the main objective of offering the reader an intuitive understanding of the rate ratio parameters in substitution models. Some evolutionary controversies were outlined, and possible solutions illustrated, to stimulate and encourage the reader to find his or her own answers. Chapters 17-22 guide the reader along a smooth path to some more advanced topics in molecular data analysis, including phylogenetic reconstruction, testing alternative phylogenetic hypotheses, and fitting discrete and continuous probability distributions to substitution data.

Two thirds of the book is suitable for an advanced undergraduate course in molecular biology and evolution, and one third ranges from the level of a graduate course to that of a professional reference. The book offers students the opportunity of deriving basic concepts and principles of molecular biology, ecology, and evolution from actual data analysis. It guides students to make their own discoveries and build their own conceptual framework of the rapidly expanding interdisciplinary science. In short, the material is developed in the spirit of the student-centered learning which is now gaining acceptance and popularity in universities around the world.

We teachers typically would try to convince our students that the teaching materials they receive from us are the best they could ever find, much in the same way as a merchant selling a spade. A spade-selling merchant will not tell us that the spade he sells is good for digging our own graves. Instead, he would try to persuade us into believing that there are treasures hidden somewhere, that the spade is a handy tool for digging up the treasure, that almost everyone has already acquired a spade, and that we would be at a terrible disadvantage if we do not acquire a spade quickly. Now to demonstrate the salesmanship that I have acquired during the last 20 years in various universities, let me share with you the secret that there is indeed much treasure hidden in large databases like GenBank, that computer programs are indeed handy tools for digging up the treasure, that almost everyone has already been using these computer programs, and that you would be at a terrible disadvantage if you fail to acquire such programs or the efficiency in using them, especially if you are going to be a student in molecular biology, ecology, and evolution.

The unique combination of the book and the computer program will allow biologists to not only understand the rationale underlying a variety of computational tools in molecular biology and evolution, but also gain instant access to these computational tools. Most of the difficult concepts were illustrated with concrete examples, and a great deal of effort has been taken to minimize the need for abstract reasoning. If you happen to belong to the unfortunate category of lesser folks who, like me, cannot see the beauty of equations without rendering them to numbers, then you may find this book exactly what you have been looking for.

Chapter 1

Installation of DAMBE and a Quick Start

DAMBE (***D**ata **A**nalysis in **M**olecular **B**iology and **E**volution*) is an integrated software package for retrieving, converting, manipulating, aligning, statistically and graphically describing and analyzing molecular sequence data, on the user-friendly Windows 95/98/NT platform. The software package has been improved dramatically since its first release in February, 1999. Extensive statistical tests of phylogenetic hypotheses have since been added, and network accessing has been much enhanced for directly accessing GenBank files or files on your networked workstations such as UNIX or Macintosh.

This chapter shows how to install DAMBE and how to get a jump start. If you have already installed DAMBE and encountered no problem, then just skip the first section and proceed to the second. Subsequent chapters will introduce more advanced techniques in descriptive and comparative analyses of molecular sequences by using DAMBE.

1. INSTALLATION

1) Download DAMBE in self-extracting format (i.e., DAMBE.EXE) from the web site: http://web.hku.hk/~xxia/software/software.htm, and save it in a temporary directory, e.g., C:\TMP.
2) Run DAMBE.EXE either in the DOS window or by clicking **Start|Run** and type in the complete path for DAMBE.EXE, e.g., C:\TMP\DAMBE.EXE in the dialog box. You will be prompted to specify which directory you wish to save the unzipped files. Just type in any directory, e.g., C:\TMP. Three files will be extracted: setup.EXE, dambe.CAB, and setup.lst. These three files can also be downloaded directly from the web page shown above.

3) If you have installed a previous version of DAMBE, I suggest that you first uninstall DAMBE before installing the new version. Click **Start|Settings|Control Panel**, and then click the **Add/Remove Programs** icon. Under the **Install/Uninstall** tag, you will find DAMBE. Click to highlight it, and then click **Add/Remove** button. Follow the prompt to completely remove DAMBE. If you have created additional files in the DAMBE directory, then these files will not be removed, and the uninstallation program will say that DAMBE is not completely removed. This is OK. If you do not see DAMBE in the **Install/Uninstall** dialog box, then you have already installed a recent version of DAMBE that does not require the uninstallation step.
4) To install, click **Start|Run**, and type in, e.g., C:\TMP\setup.exe. Follow the instructions on the screen and use the default whenever possible. The installation is of two parts. First, DAMBE will check to see if you may have outdated Microsoft DLL and EXE files. If you do not, then the first part is omitted and DAMBE installation will start right away. If you have outdated system files, however, DAMBE will first prepare your system to update these programs. You may be asked to restart your computer. Click **Yes**. After restarting your computer, the first part is completed. One thing that occasionally confuses the user after restarting is that the computer does not automatically go to the second part of the installation.You should now click **Start|Run** and again type in C:\TMP\setup.exe. This time DAMBE will omit the first part of installation and proceed to the second part of DAMBE installation. Follow the instructions on the screen and use the default whenever possible. The default installation directory is c:\Program Files\DAMBE.
5) After installation, a program icon will be added to the Start menu. You may now run the program from the Windows desktop by clicking **Start|Dambe*.*|DAMBE**, where *.* stands for the version number of DAMBE.

2. A JUMP START

After the installation, you will find a number of data files in the directory where DAMBE.EXE resides. These data files are for you to practice with DAMBE, but it would be better if you have your own data files in some of your directories. The various file formats represented by the sample files may be confusing at first, and you should ignore them for the time being. Chapter 2 provides an introduction to the plethora of file formats, the rationale underlying these various file formats, and how to use DAMBE to convert these formats into each other.

You can now start the program by clicking the program icon from the program start menu. A standard window interface appears (fig. 1), waiting for your input. The display window will automatically show scroll bars when there are more text than can be displayed in the window.

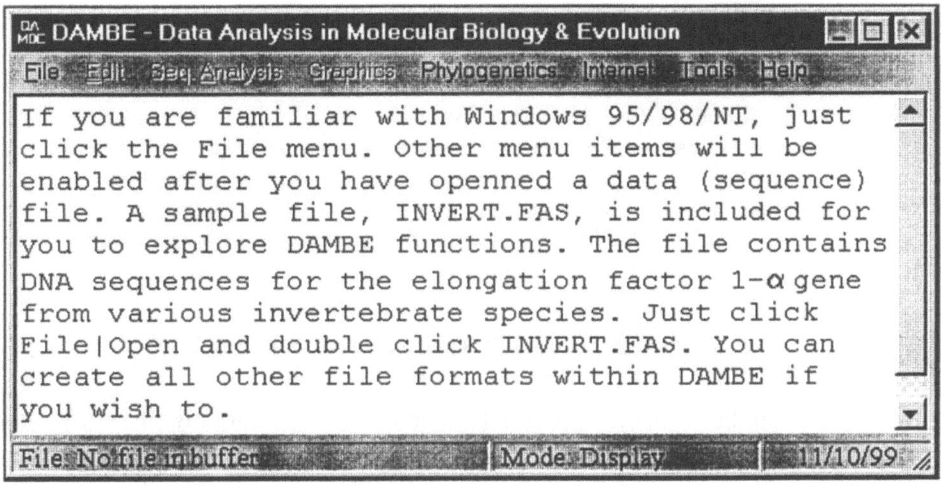

Figure 1. DAMBE's starting window, which is also used for displaying text output.

Click the **File** menu, then click the **Open** menu item (which will be abbreviated as **File|Open** in subsequent chapters). The standard WINDOWS **file/open** dialog box appears (fig. 2). This dialog box is used in DAMBE for all file input/output. Note that, by default, only files with .FAS extension are shown, to avoid cluttering of the screen. If you click the **Files of Type** dropdown listbox and select another file type, say MEGA files, then only files with file extension .MEG will be shown. For the time being, just leave the file type as .FAS. Double-click the file **INVERT.FAS**, which contains seven nucleotide sequences of the elongation factor 1-α gene from seven invertebrate species. Alternatively, you can click the file once to highlight it, and then click the **OPEN** button.

This standard **file/open** dialog box can perform some simple file management tasks. For example, if you want to delete a file, just right-click your mouse and then click **delete** in the pop-up menu, and the file will be deleted to the wastebasket. If you wish to delete the file completely, then hold down the shift key and then click delete. If you wish to change a file name, just click the file to highlight it, and then click it once more. Now you can just type in the new file name. But please do not delete any file in the DAMBE directory or change any file name.

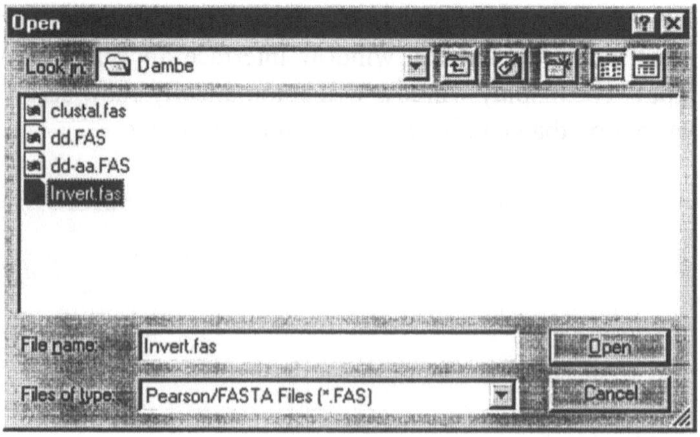

Figure 2. Standard **file open** dialog box in DAMBE

After you have opened a file (either by double-clicking it or by first highlighting it and then clicking the **Open** button), a dialog box appears requesting the nature of the sequences (fig. 3), i.e., whether the input file contains non-protein-coding sequences (e.g., rRNA sequences), amino acid sequences or protein-coding nucleotide sequences. The reason for DAMBE to request this information is because different types of sequences are often associated with different analytical methods. DAMBE will make different analytical options available according to the type of input sequences.

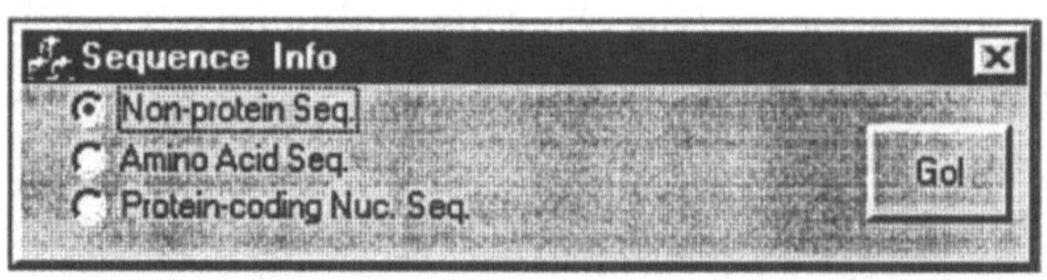

Figure 3. Type of molecular sequences

If your sequences are protein-coding nucleotide sequences, as are the sequences in the **invert.fas** file, then you should click the option for protein-coding sequences. Because different organisms may use different genetic codes to translate mRNA molecules to proteins, DAMBE will present another set of options for you to choose which genetic code is associated with your protein-coding sequences, i.e., whether it is universal or mammalian mitochondrial or any of the other ten genetic codes (fig. 4). Click the appropriate radio button, and then click **Go!**. If the sequences are not aligned, then you will be asked whether you wish to aligned the sequences. The sequences are then shown in the display window, and are now stored in the computer memory waiting for you to apply analyses to them. Do whatever you consider sensible, otherwise please proceed to read the next chapter, or just click **File|Exit** for now and come back later (**File|Exit** means that you first click the **File** menu and then click the **Exit** item).

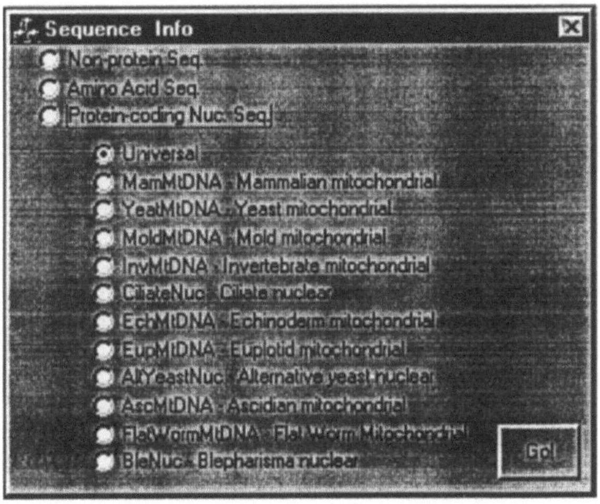

Figure 4. Choosing among genetic codes

Chapter 2

File Conversion

Molecular data come in many different formats, some of which are represented by sample files that come with DAMBE. These sample files are located in the directory where DAMBE.EXE resides. If you have already used PHYLIP and PAUP, then you already know at least two file formats and the difference between them. If you have retrieved sequences from GenBank, you might have already noted the difference between the GenBank format (one of the most complicated sequence formats) and the FASTA format (one of the simplest sequence formats), which are the only two formats in which GenBank delivers the sequences to your networked computer. Sequences in the PHYLIP or PAUP formats are **aligned**, and are typically represented in **interleaved** format. Sequences in the GenBank format are not aligned and are represented in **sequential** format. Sequences in FASTA format can either be aligned or not aligned, and are represented in **sequential** format. One should use interleaved format to represent aligned sequences.

If you have not encountered any of these file formats, then it is now a good time to have a look at these files, all of which are plain text files. There is an ugly but convenient built-in file viewer in DAMBE under the **Tools** menu which you can use to view most text or graphics files. These sample files are provided in case you have not yet engaged in any real data analysis in molecular evolution and phylogenetics, and consequently have not accumulated a private collection of data files.

If you have wondered why DAMBE should support so many different file formats, here is the answer. Although DAMBE covers a substantial amount of computational tools used in molecular biology and evolution, many users will certainly find other special-purpose programs with functions not available in DAMBE. Many of these special-purpose programs use nucleotide or amino acid sequence files with special (or even weird) input formats. For this reason, DAMBE provides you with an extensive file conversion utility to facilitate your data analysis with other programs.

This chapter will first bring you into contact with a plethora of commonly used computer programs used in bioinformatics and molecular biology and evolution, and the commonly used sequence formats associated with these computer programs. It will then introduce you to one of the commonly used file conversion utility, READSEQ, and outline some of its limitations. Finally, you will learn how to convert files between different file formats using DAMBE.

Two file conversion utilities are available in DAMBE, one converting all sequences in a file from one format to another, and the other converting a subset of sequences in your file from one format to another. You can also convert protein-coding nucleotide sequences in one format into amino acid sequences in another format.

1. A PLETHORA OF COMPUTER PROGRAMS

Scientists in the field of molecular biology and evolution use a variety of computer programs, with functions covering comparative sequence analysis, sequence alignment, protein and RNA structure, gene identification, data mining, and so on. You should learn to take advantage of the power of these programs in carrying out data analysis of molecular data. Most programs were written by active researchers who wish to solve one specialized problem in their own research but then feel that the resulting program might be useful to others as well. The following URL's list computer programs commonly used in data analysis in molecular biology and evolution, as well as links to other software listings:

http://evolution.genetics.washington.edu/phylip/software.html
http://biosci.biosc.lsu.edu/general/software.html
http://darwin.eeb.uconn.edu/molecular-evolution.html
http://www.york.biosis.org/zrdocs/zoolinfo/software.htm
http://www-biol.univ-mrs.fr/english/ftp.html
http://iubio.bio.indiana.edu:81/soft/biosoft-catalog/

2. A PLETHORA OF SEQUENCE FORMATS

The plethora of computer programs results in a plethora of file formats. There are currently 18 file formats in common use in molecular biology and evolution, and I hope that the number will become stabilized. These 18 formats, together with what DAMBE can read in and convert to, are listed below. It is good practice to associate each file format with one particular file type. If you have used Microsoft Office, you will notice that WORD files are associated with the .DOC file type, EXCEL files with the .XLS file type, and PowerPoint files with the .PPT file type.

Table 1. Common data file formats used in DAMBE. The last two columns specify what DAMBE can do.

Sequence Format	File Type	Read in	Convert to
Phylip	.PHY	Yes	Yes
PAUP	.PAU	Yes	Yes
MEGA	.MEG	Yes	Yes
CLUSTALW	.ALN	Yes	
FASTA	.FAS	Yes	Yes
GenBank	.GB	Yes	Yes
GCG	.GCG	Yes	Yes
MSF	.MSF	Yes	Yes
DNAStrider	.STR	Yes	Yes
PAML	.PML, .NUC	Yes	Yes
RST,MP	.RST, .MP	Yes	
PHYLTEST	.SK		Yes
IG/Stanford	.IG	Yes	Yes
NBRF	.NBR	Yes	Yes
EMBL	.EMB	Yes	Yes
Fitch	.FIT	Yes	Yes
PIR/CODATA	.PIR	Yes	Yes
Allele Frequency	.FRE	YES	

If you hate to read this chapter, or confused by the preponderance of file formats, then try to persuade programmers not to create more file formats. Don Gilbert has made this appeal a long time ago, unfortunately without much effect.

3. READSEQ

READSEQ is an excellent program written by Don Gilbert, and can automatically recognize and convert many file formats into each other. I personally have benefited greatly from using the excellent yet free program. However, it has five major limitations:

1. READSEQ cannot read or write the following sequence formats that can be processed by DAMBE:
   - MEGA: sequential and interleaved formats
   - PAML: sequential and interleaved formats, and the RST format which contains a tree structure and the reconstructed ancestral sequences,

generated in PAML or DAMBE when the user chooses to reconstruct ancestral sequences using the maximum likelihood method (Yang et al. 1995)
  - CLUSTAL: the aligned sequences
  - PHYLTEST: a very special format that is easy to output with DAMBE.
2. READSEQ does poorly with GenBank files, which contains a lot of information (e.g., beginning and ending sites of a coding sequence, an intron, an exon, a rRNA sequence, etc) about the sequences. READSEQ simply ignores all this information and read in the whole sequence. In contrast, when DAMBE reads in a GenBank file, it automatically takes in all these pieces of information and allows you to splice out the desired sequence segments. See the chapter entitled "PROCESSING GENBANK FILES" for details.
3. READSEQ, being a text-based program, is clumsy at saving a subset of sequences. In contrast, DAMBE allows you to list all sequences and simply click a subset of sequences for saving into any specified file format.
4. READSEQ does not read in long sequence names in several formats, resulting in truncation of sequence names.
5. READSEQ is slow when reading large sequence files.

4. FILE CONVERSION USING DAMBE

DAMBE provides two convenient ways for you to convert your sequence files from one format to another. The first allows you to convert all the sequences, and the second allows you to save a subset of sequences in your file. The latter is useful in the following situations:

- You wish to do a phylogenetic analysis, but the phylogenetic program complains that there are too many sequences in your file. Some phylogenetic programs, such as CODEML in the PAML package, are very slow and simply cannot deal practically with more than 10 sequences.
- The sequences in your file is heterogeneous, e.g., contain sequences for two or more different genes. This is particularly true when you retrieve sequences from GenBank by searching with keywords. You consequently may wish to save them into different files, each containing orthologous sequences for one gene.

The input sequences for DAMBE may contain characters such as "-", "?" and ".", which are interpreted, respectively, as a gap, an unresolved base, and a base identical to the first sequence at the same site. All saved files are plain text files. All occurrences of T are changed to U in the computer buffer.

4.1 Convert all sequences from one format to another

Start DAMBE, and open a sequence file according to the instruction close to the end of the last chapter. The sequences will be displayed in the display window. Click **File|Save As (Converting sequence format)**. The standard **file/open** dialog box appears. Choose the appropriate file format and click OK. You will be informed that the file has been saved into a text file. Click OK, and the converted file will be shown on the screen (so that you are sure of the correctness of the conversion). You see that the program is very user-friendly. This is true also when you perform more complex data manipulation and analyses using DAMBE.

Here are some particulars pertaining to some formats:

MEGA: MEGA file format allows some comments. You will be prompted to enter a description.

PIR: PIR format is for amino acidsequences. If the sequences you are converting are nucleotide sequences, you will be informed that the PIR format is for protein sequences and prompted as to whether you want to translate the nucleotide sequences into amino acid sequences. In the latter case, the user needs to tell DAMBE at which nucleotide site to begin translation. This is necessary for the following reason. Take the following nucleotide sequence GCU GGU AUG U for example. The resulting amino acid sequence is Ala-Gly-Met if DAMBE starts translation from the first nucleotide site (the trailing partial codon represented by U is ignored). However, the sequence would be translated to Leu-Val-Cys if DAMBE starts translation at the second nucleotide site. PIR output is in single-letter notation, i.e., each amino acid is represented by a single letter.

GCG: There are two file formats in GCG, the single file format with file extension .GCG, and the multi-sequence file format with the file extension .MSF. If your original sequence file contains multiple sequences and you choose the file type .GCG, you will be asked whether you actually wish to save the sequences into the multi-sequence format. If you choose **Yes**, then the file, with multiple sequences, will be saved in GCG MSF format, otherwise the sequences will be saved to the file in GCG single sequence format.

4.2 Converting a subset of sequences

Start DAMBE, and open a sequence file if you have not done so already. The sequences will be displayed in the display window. Now click **File|Save a subset of sequences**. A dialog box appears for sequence selection (fig. 1). A similar dialog box (or slight variation of it) will also appear when you choose sequences for other types of manipulation or analysis. It is therefore worthwhile to pause a minute to get familiar with this dialog box.

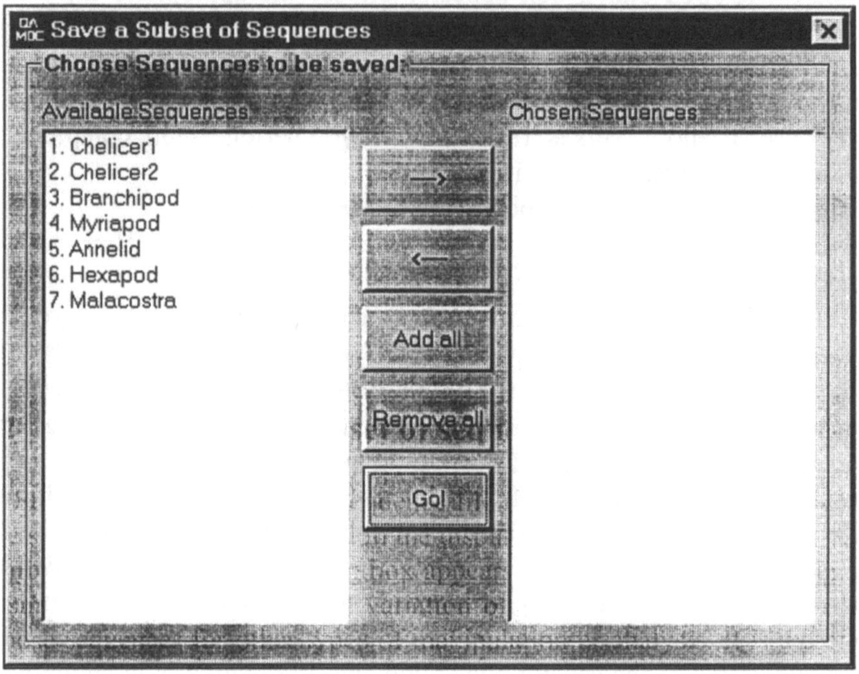

Figure 1. Dialog box for file selection

There are two lists in the dialog box. The one on the left shows the sequences that are available for selection. The one on the right displays sequences selected for output. At this moment, the list on the right is empty.

- To select a single sequence, just click to highlight it, and then click the → button to move it to the right. If you have made a mistake and transferred a wrong sequence to the right, then just click to highlight the sequence and click the ← button to move it back to the left.
- To select neighboring sequences, click the first of the neighboring sequences to highlight it and then, while holding down the shift key, click the last of the neighboring sequences. All the neighboring sequences will then be highlighted. Click the → button to move them to the right.

– To select disjoint sequences, click each sequence while holding down the Ctrl key, and then click the → button to move the highlighted sequences to the right.

This selection procedure is the same when you perform other functions involving sequence selection. Once you have finished your selection, click the **Go!** button. A standard **file/save** dialog box appears. Choose the desired file type (sequence format). Type in the file name for saving the result, or simply use the default. Then click the **Save** button. The file is saved in text format, and also displayed in the display window (to assure you of the correctness of the conversion).

You can translate any protein-coding nucleotide sequences into amino acid sequences by using any one of the 12 implemented genetic codes. Translation depends on which genetic code you use. All 12 known genetic codes have been implemented in DAMBE (Details of these genetic codes are listed in http://web.hku.hk/~xxia/software/GenCode.htm):

1. universal
2. vertebrate mitochondrial
3. yeast mitochondrial
4. mold mitochondrial
5. invertebrate mitochondrial
6. ciliate nuclear
7. echinoderm mitochondrial
8. euplotid mitochondrial
9. alternative yeast nuclear
10. ascidian mitochondrial
11. flat worm mitochondrial
12. blepharisma nuclear

4.3 Output PHYLTEST files

You might want to skip the rest of the chapter if you do not use PHYLTEST written by Sudhir Kumar. The program is primarily developed to facilitate the use of statistical tests of phylogenetic hypotheses based on the minimum evolution (ME) principle. For further theoretical considerations and for mathematical formulae, you may refer to relevant literature for the ME method (Rzhetsky et al. 1995; Rzhetsky and Nei 1992a; Rzhetsky and Nei 1992b; Rzhetsky and Nei 1993).

PHYLTEST can take nucleotide sequences, amino acid sequences, or a distance matrix as input. The file format involving nucleotide sequences is rather complicated, but can be easily generated by using DAMBE. All descriptions below pertain to molecular sequence data.

4.3.1 A PHYLTEST sample file

```
12S rRNA data from Cooper et al.
nucleotide
13 370

#emu_{emu}
GCTTAGCCCTAAATCTTGATACTCACCTTACCAGAGCATCCGCCTGAGAACTACGAGCACAA
ACGCTTAAAACTCTAAGGACTTGGCGGTGCCCTAAACCCACCTAGAGGAGCCTGTTCTATAA
TCGATAACCCACGATACACCCAGCCATCTCTTGCCACAGCCTACATACCGCCGTCGCCAGCC
CGCCTATGAAAGATAGCGAGCACAATAGCCCGCTAACAAGACAGGTCAAGGTATAGCGTATG
AGATGGAAGAAATGGGCTACATTTTCTAACATAGAATAACGAAAGAAGATGTGAAATCCTTC
AGAAGGCGGATTTAGCAGTAAAACAGAATAAGAGAGTCTATTTTAAACTGGCTCTAGGGC

#cassowary_{cassowary}
GCTTAGCCCTAAATCTTGATACTCGCTATACCTGAGTATCCGCCCGAGAACTACGAGCACAA
ACGCTTAAAACTCTAAGGACTTGGCGGTGCCCTAAACCCACCTAGAGGAGCCTGTTCTATAA
TCGATAACCCACGATACACCCAACCATCTCTTGCCACAGCCTACATACCGCCGTCGCCAGCC
CGCCTGTGAGAGATAGCGAGCATAACAGCCCGCTAACAAGACAGGTCAAGGTATAGCGTATG
AGATGGAAGAAATGGGCTACATTTTCTAACATAGAATAACGAAAAAGGATGTGAAATTCCTT
AGAAGGCGGATTTAGCAGTAAAACAGAACAAGAGAGTCTATTTTAAACCGGCCCTAGGGC

#kiwi1_{kiwi}
GCTTAGCCCTAAATCCTGGTACTTACGTTACCTAAGTACCCGCCCGAGAACTACGAGCACAA
ACGCTTAAAACTCTAAGGACTTGGCGGTGCCCTAAACCCACCTAGAGGAGCCTGTTCTATAA
TCGATAACCCACGATACACCCAACCATCTCTTGCCACAGCCTATATACCGCCGTCGCCAGCT
CGCCTATGAGAGACAGCGAACACAACAGCTAGCTAACAAGACAGGTCAAGGTATAGCCTATG
AGATGGAAGAAATGGGCTACATTTTCTAAAATAGAATAACGAAAAAGGGTGTGAAATCCCTT
AGAAGGCGGATTTAGCAGTAAAACAGAATAAGAGAGTCTATTTTAAGCTGGCCCTAGGGC
#kiwi2_{kiwi}
GCTTAGCCCTAAATCCTGGTGCTTACATTACCTAAGTACCCGCCCGAGAACTACGAGCACAA
ACGCTTAAAACTCTAAGGACTTGGCGGTGCCCTAAACCCACCTAGAGGAGCCTGTTCTATAA
TCGATAACCCACGATACACCCAACCATCTCTTGCCACAGCCTATATACCGCCGTCGCCAGCT
CGCCTATGAGAGACAGCGAACACAACAGCTAGCTAACAAGACAGGTCAAGGTATAGCCTATG
AGATGGAAGAAATGGGCTACATTTTCTAAAATAGAATAACGAAAAAGGGTGTGAAATCCCTT
AGAAGGCGGATTTAGCAGTAAAACAGAATAAGAGAGTCTATTTTAAGCTGGCCCTAGGGC

#rhea1_{rhea}
GCTTAGCCCTAAATCCTGATACTTACCCCACCTAAGTATCCGCCCGAGAACTACGAGCACAA
ACGCTTAAAACTCTAAGGACTTGGCGGTGCCCTAAACCCACCTAGAGGAGCCTGTTCTATAA
TCGATAACCCACGATACACCCGACCATCTCTTGCCCCAGCCTACATACCGCCGTCCCCAGCC
CGCCTGTGAAAGACAGCAGGCATAATAGCTCGCTAACAAGACAGGTCAAGGTATAGCATATG
GGATGGAAGAAATGGGCTACATTTTCTAATCTAGAACAACGGAAGAGGGCATGAAACCCCTC
CGAAGGCGGATTTAGCAGTAAAGTAGGATCAGAAAGCCCACTTTAAGCCGGCCCTAGGGC
#rhea2_{rhea}
GCTTAGCCCTAAATCCCGATACTTACCCCACCCAAGTATCCGCCCGAGAACTACGAGCACAA
```

```
ACGCTTAAAACTCTAAGGACTTGGCGGTGCCCTAAACCCACCTAGAGGAGCCTGTTCTATAA
TCGATAACCCACGATACACCCGACCATCTCTTGCCCCAGCCTACATACCGCCGTCCCCAGCC
CGCCTATGAGAGACAGCAAGCATAATAGCTCGCTAGCAAGACAGGTCAAGGTATAGCATATG
AGATGGAAGAAATGGGCTACATTTTCTAGTCTAGAACAACGAAAGAGGGCATGAAACCCCTC
CGAAGGCGGATTTAGCAGTAAAGTGGGATCAGAAAGCCCACTTTAAGCCGGCCCTAGGGC
```

4.3.2 Generating PHYLTEST files with DAMBE

Start DAMBE and read in a sequence file. Click **File|Save As**, and a standard **File/Save** dialog box will show up. Click the **Save as type** dropdown menu, and choose the PHYLTEST file type (the second last in the dropdown list). A dialog box is then displayed (fig. 2). Click a set of sequences that you know are monophyletic and then click the → button to move them to the right. Now enter a one-word ID for the group and click the **Done** button. Continue this process until all sequences have been processed. The finished file will be automatically displayed in the display window to assure you of the correctness of the conversion.

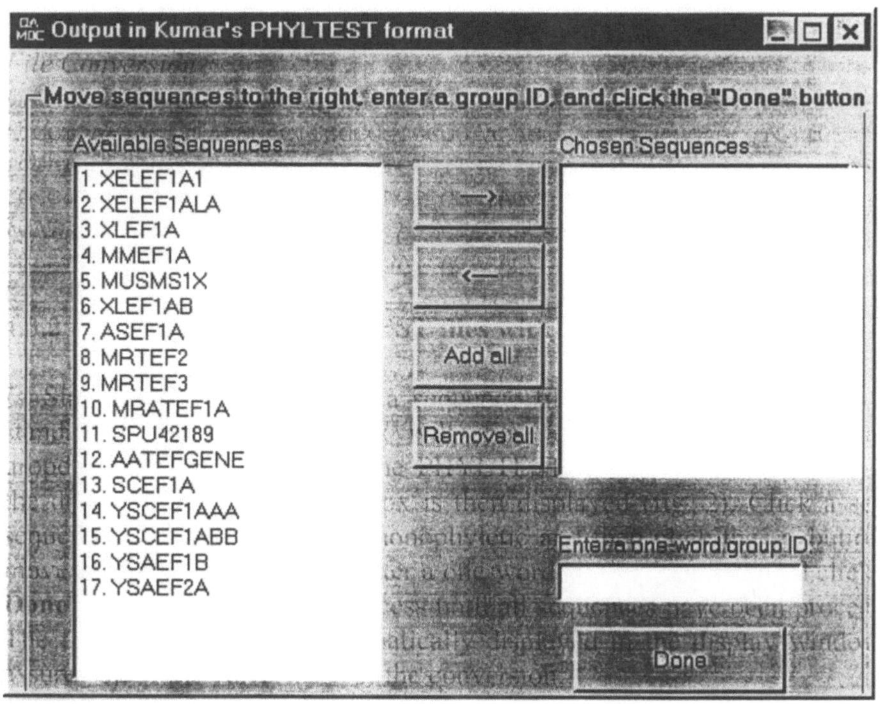

Figure 2. Dialog box for generating the PHYLTEST input file

Chapter 3

Processing GenBank Files

If you ask an expert in bioinformatics what is the most important resource in the modern world, he will most likely give you a surprising answer. He will tell you that the most important resources are not whales in the ocean, or minerals on land or petroleum underground. The most important resource, he will argue, lies in molecular databanks such as GenBank. What modern people should do is not to make giant ocean fleets to kill those already threatened or endangered marine species, neither should they drill deep underground to take up the already depleted petroleum reserves. What modern people should be doing is to design efficient software to get at the treasures hidden in those large and ever-expanding molecular databanks.

The wisdom in the assertion by the bioinformatics expert may not be immediately obvious to you. However, it is my belief that you will very soon be making the same assertion, and will find GenBank a part of your life.

DAMBE allows you to read molecular sequences directly from GenBank if your computer is connected to internet. This function has been handy and time-saving for me. For example, if I come across a paper that listed a number of protein-coding sequences with either GenBank accession numbers or LOCUS names, and I want to verify the claims made by the author(s), all I need to do is simply click **File|Read sequences from GenBank** and type in the accession numbers or LOCUS names. DAMBE will splice out the introns and join the CDS automatically by taking advantage of the FEATURES table in the GenBank sequence file, align the sequences and allow me to carry out comparative sequence analyses with no hassle.

The power of DAMBE will be better appreciated if you know something basic about GenBank sequence format and how the information is stored in GenBank files.

Sequence files in GenBank can be retrieved in one of two formats via Internet. One format is the FASTA format, which is one of the simplest sequence formats, and the other is the GenBank format, which is one of the most complicated sequence formats. These two file formats can both be directly read into DAMBE. Sequence files in the FASTA format contain just plain sequences as well as sequence names to designate each of the sequences. The sample file **invert.fas** is a typical sequence file in FASTA format.

The GenBank format, designated by the file type .GB in DAMBE, features rich annotations for the molecular sequence. Each sequence in the file has a LOCUS name, and may have one or more accession numbers. Each of the sequences may contain multiple coding regions (CDS), multiple introns and exons, and multiple rRNA genes. These different segments within the same sequence are specified in what is known as the FEATURES table in GenBank files.

Because of the complexity of the GB files and the frequent necessity of utilizing the rich information contained in GB files, I have written this chapter entirely on how to deal with GB files. You will first learn some basics about the FEATURES table of a typical GB file, and then learn how to use DAMBE to read in GB files while taking advantage of the information contained in the FEATURES table. You may skip this chapter if you are not going to work with GB files in the near future.

1. GENBANK FILE FORMAT

A typical, but abridged, GenBank file, which contains the elongation factor-1α (EF-1α), is shown below. The complete file can be found in the file EF1A.GB in the installation directory of DAMBE. GB files are plain text files which you can view within DAMBE by using the built-in file viewer under the **Tools** menu.

```
LOCUS       MRTEF2        2263 bp    DNA              PLN       17-FEB-1997
DEFINITION  Mucor racemosus TEF-2 gene for elongation factor 1-alpha
ACCESSION   X17476
......
FEATURES             Location/Qualifiers
     source          1..2263
                     /organism="Rhizomucor racemosus"
     ......
     CDS             join(464..517,646..1735,1933..2165)
                     /codon_start=1
                     /product="EF-1-alpha"
     ......
     exon            <464..517
```

```
                     /number=1
     intron          518..645
                     /number=1
     exon            646..1735
                     /number=2
     ......
     intron          1736..1932
                     /number=2
     exon            1933..>2165
                     /number=3
BASE COUNT      572 a    511 c    480 g    700 t
ORIGIN
        1 tttttctcat tgggaatcca ttggaatgaa aggacaaatg cactctcgca atgagatgct
       61 ttaaatgctg gcaaatttga aggatgtaca atcgaaactt tccaaatgtc ctcaaacaag
     ......
     2161 aataaattgc tacatagtag ttttttcttt cccattgctg tcagtatata gtaaaagccc
     2221 ttgtacagtg tgctttggat ttaaattatt caaaataaat caa
//

LOCUS       MRTEF3        1881 bp    DNA              PLN       10-APR-1993
DEFINITION  Mucor racemosus TEF-3 gene for elongation factor 1-alpha
ACCESSION   X17475
......
FEATURES             Location/Qualifiers
     ......
     CDS             join(88..141,200..1289,1490..1719)
                     /codon_start=1
                     /product="EF-1-alpha"
     ......
     exon            <88..141
                     /number=1
     intron          142..199
                     /number=1
     exon            200..1289
                     /number=2
     ......
     intron          1290..1489
                     /number=2
     ......
     exon            1490..>1719
                     /number=3
BASE COUNT      459 a    436 c    413 g    573 t
ORIGIN
        1 ggatccatcc atgccacaaa tcagcataaa tgctatccat ccatccatca aacatactta
       61 catgtatcat ctttcattat agtcgcaatg ggtaaggaga agactcacgt taacgtcgtc
     ......
```

```
     1801 ataatctgta taagttgtgt tgtccatgac gtgatgtgag gtgtgtttat tgagtggtgc
     1861 acatcagttc gtatattagg a
//

LOCUS ......
```

Every molecular sequence in GenBank is assigned a LOCUS name, e.g., MRTEF2 is the LOCUS name for the first DNA sequence in the GB file shown above. It contains a nucleotide sequence with 2263 bases, which are numbered from 1 to 2263. Notice that the EF-1α genes in the two sequences each contain three exons, and the final coding mRNA results from the splicing out of the introns and the joining of these three exons. The CDS entry in the FEATURES table specifies the location of these three coding segments, with the first starting and ending at positions 464 and 517, respectively, the second starting and ending at positions 646 and 1735, respectively, and so on. The mature mRNA, i.e., the mRNA exported to the cytoplasm to specify the translation into the amino acid sequence, results from the joining of these three segments.

For those of us who study molecular biology and evolution, it is often necessary to splice out a particular DNA sequence from a variety of species and make interspecific comparisons. For example, to study the evolution or functional changes of the coding sequences of the elongation factor-1α, it is necessary to splice out the CDS regions of EF-1α and join them together, and repeat this process for a variety of organisms in order to make interspecific comparisons. Similarly, to study the evolution of introns of EF-1α, one would need to splice out the introns from a variety of organisms and make comparisons among them. To cut out and join these different sequence segments manually or with the aid of a text editor would be very cumbersome and error-prone. DAMBE fully automate the whole process in an elegant and pleasing way. What you need is just a few simple clicks of a mouse button.

2. REANDING GENBANK FILES WITH DAMBE

The best way to proceed now is to run DAMBE and see how it works. Start DAMBE and click **File|Open**. A standard file dialog box appears. Go to the installation directory of DAMBE where the **EF-1A.GB** file is located. It should be in the directory C:\Program Files\DAMBE if you installed the program by default. In the **File of type** dropdown listbox, choose (click) GenBank file format. You will see **EF-1A.GB** file in the dialog box. Double-click it, or single-click it to highlight it and then click the **Open** button. A dialog box appears (fig. 1), prompting you to choose whether to read in the whole sequence or specific segments within each sequence specified in the FEATURES table in the GenBank file. Occasionally you may have GenBank files that do not have the FEATURES table, in which case you should choose the default, i.e., reading the whole sequence. Note that some GenBank sequences may take several megabytes of space and you should be cautious about reading in the whole sequence. If the GenBank file contains amino acid sequences, then you should click the last option, i.e., **Amino acid sequence**.

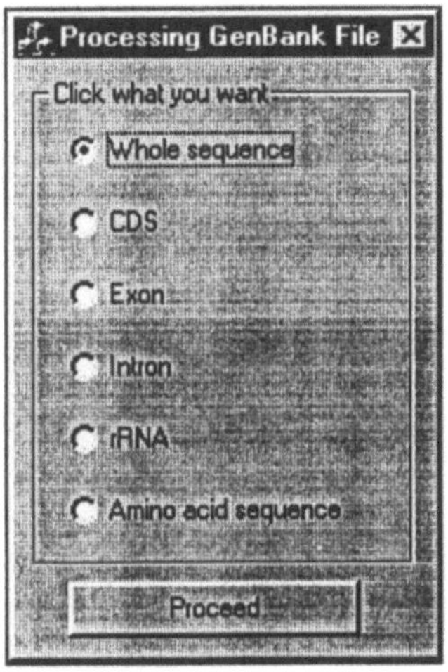

Figure 1. Dialog box for reading sequences in GenBank format

If you choose to read in the whole sequence (the first option), or if the input file contains amino acid sequences only (the last option), then the sequences in the GenBank file will be read in sequentially, with the LOCUS name used as the sequence name. If your input file contains nucleotide sequences with a FEATURES table specifying the nature of individual segments (e.g., CDS, exon, intron, rRNA, etc.), then you can choose to read in particular segments from each sequence.

For practice, let's assume that you wish to get the coding sequences (CDS) specifying the EF-1α protein from the two nucleotide sequences contained in the file EF1A.GB. Click the **CDS** button and then click the **Proceed** button, Another interactive dialog appears and is partially shown in fig. 2. There are five list boxes, with two listboxes not shown in fig. 2. The first column shows the LOCUS name of each GenBank sequence, the second shows the length of each sequence, and the third is taken from the DEFINITION entry of the GenBank sequence. The fourth and the fifth list boxes are currently empty. What you wish to get out of the GenBank file is specified under **Splice**, which is CDS for this operation.

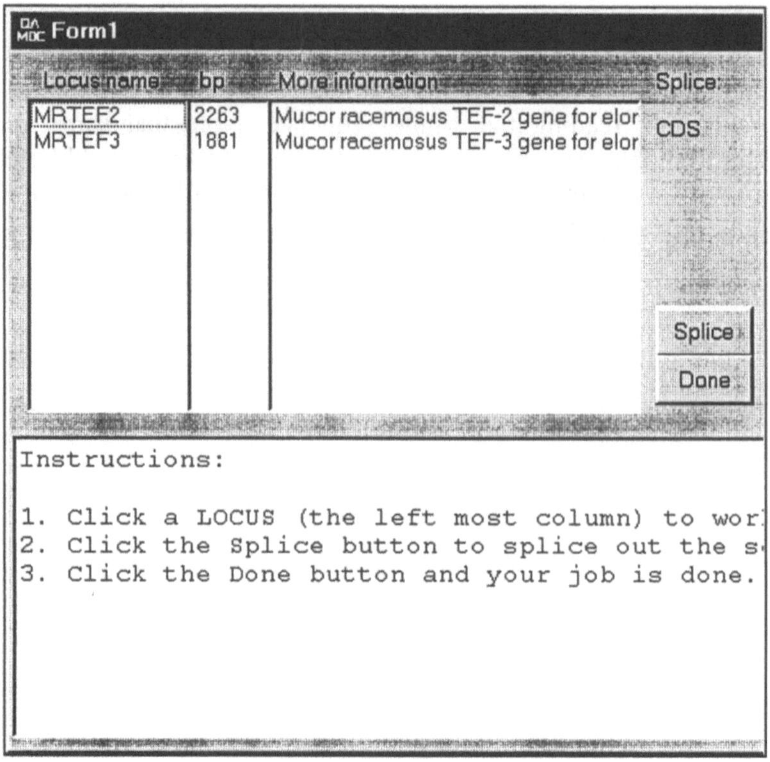

Figure 2. Reading GenBank file by taking advantage of the information contained in the FEATURES table. The dialog box is too wide to be shown in its entirety and only the left half is shown.

There are also some hidden boxes. For example, some sequences were deposited as complementary strand, and the GenBank file will state so in the FEATURES table. DAMBE will take this information and automatically get the correct opposite strand, i.e, the actually transcribed RNA sequence. In this case, a text box with the word COMPLEMENT will be displayed. Because our sequences are not the complementary sequence, this text box will remain hidden.

Now click the first LOCUS name, i.e., MRTEF2. The dialog box will change to display sequence-specific information for the LOCUS MRTEF2 (fig. 3). The fourth list box displays the name of the target CDS sequence in MRTEF2. In our sample file EF1A.GB, there is only one CDS named EF-1alpha, whose three segments are specified in the fifth list box. Let me explain briefly the numbers on the fifth listbox. The EF-1α gene in the two Mucor species is made of several exons with introns in between. At the beginning and the end of the coding sequence there are also untranslated sequences. What we have retrieved from GenBank are two sequences with each specifying where the coding segments are located. For example, the MRTEF2 sequence is 2263 bases long, with the first coding segment beginning at position 464 and ending at 517, the second coding segment starting from 646 and ending at 1735, and the third coding segment starting from 1933 and ending at 2165. The complete coding sequence, the mature mRNA, is made by joining these three segments.

The text box in the lower panel displays the complete sequence with the three segments color-coded in red (fig. 3). You might have noticed that the first codon is ATG, which is the initiation codon, and the last codon is TAA, which is the termination codon. This means that our CDS specifies a complete protein gene.

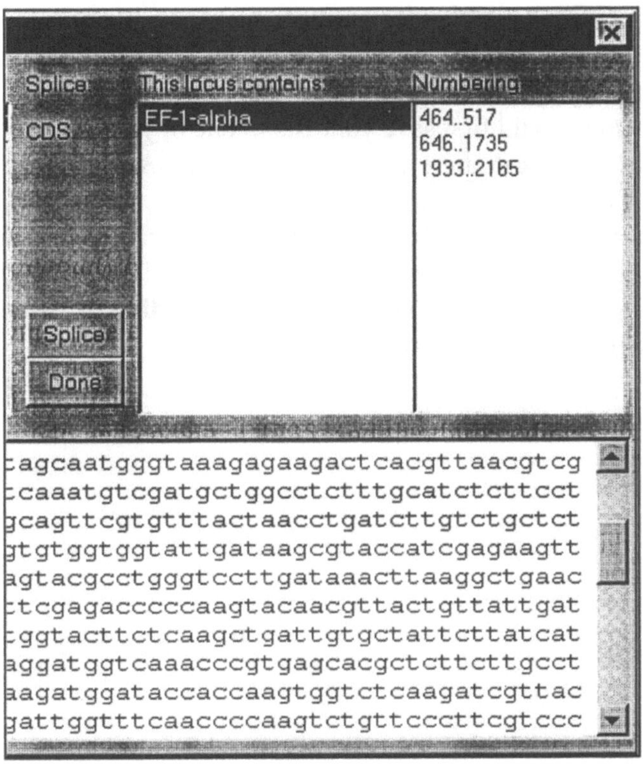

Figure 3. Dialog box displaying specifications for the coding sequence. The complete nucleotide sequence specifying the protein is made of three segments: 464..517, 646..1735, 1933..2165. The three segments were shown in the text box in the lower panel in red. Only the right half of the dialog box is shown because the dialog box is too wide.

Click the **Splice** button to splice out and join these three segments, and repeat this process for the second LOCUS, i.e., MRTEF3. There are only two

LOCUS's in the EF1A.GB file, so we have finished our operation of splicing and joining. Click the **Done** button, and you will be prompted to confirm the type of sequences, which we have encountered several times already. Just click the option button **Protein-coding Nuc. Seq.** and then choose **Universal** as the genetic code.

A bell rings, and a dialog box comes up telling you that the two CDS sequences are not of equal length, and asking if you wish to align the sequences with CLUSTALW (Thompson et al. 1994). I recommend that you click **NO** because we have not yet learned anything about how to specify the parameters for alignment. The unaligned sequences will then be shown in the display winhdow. If you are adventurous, you may click **YES** and use the default parameter specification for sequence alignment. DAMBE includes a large part of ClustalW codes for multiple sequence alignment. The multiple alignment is slow. Once the alignment is done, the aligned sequences will be shown in the display window for you to apply any analysis on them. Usually at this stage you should first save your file in one of your favourite formats.

What we have just done is to splice the CDS sequences in the two LOCUSes. You can also splice out introns, exons, rRNA, etc, in the same way. You should now start from the beginning by re-opening the EF1A.GB file and try to splice out the exons as an exercise. If you wish to do a more adventurous exercise, click **File|Read sequences directly from GenBank**, which we will cover in the next chapter.

Chapter 4

Accessing GenBank or Other Networked Computers

1. INTRODUCTION

In this chapter you will learn two skills related to internet. One is to read molecular sequences directly from GenBank, and the other is to read files of molecular sequences from, or write files to, your networked computers. The latter is useful when you want to use DAMBE to analyze your data stored on another computer, or when you want to use DAMBE to format sequences for further analysis by using special software installed on another computer. DAMBE essentially makes GenBank or your networked computer behave like another hard drive on your local Windows-based PC.

2. READING MOLECULAR SEQUENCES DIRECTLY FROM GENBANK

Start DAMBE if you have not done so. Click **File|Read Sequences from GenBank**, and a dialog box appears (Fig. 1) for specifying options. GenBank sequences can be accessed by the accession number, the LOCUS name or keywords. Consequently, you have two search methods, one by using GenBank accession number or LOCUS name or the combination of the two, and the other by using keywords. It is important to keep in mind that there are now many sequences in GenBank and a keyword search may produce a large number of hits. For example, if you use *Homo sapiens* as keywords, then you will get more than a million sequences in the current release of GenBank. Of course your hard disk will be filled up long before you could ever get that many sequences. It is for this reason that I have included an option for setting the upper limit of hits, which can range from 10 to 1000. Make sure that you formulate the keywords carefully to get what you want. An example of searching with keywords is illustrated in fig. 1. The search string tells DAMBE to retrieve the first 20 nucleotide sequences in GenBank that contain words "Geomys" and "cytochrome". "Geomys" is the generic name for a group of small rodents called gophers.

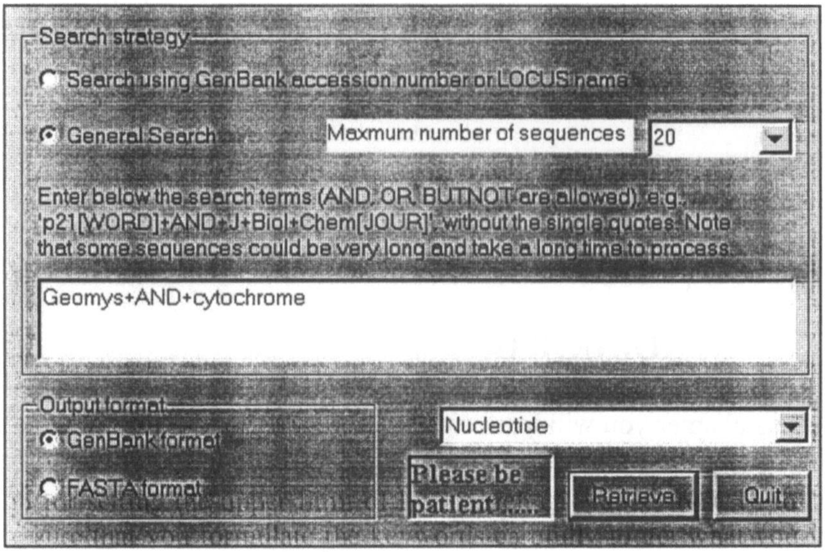

Figure 1. Specifying options for retrieving molecular sequences from GenBank

It is simpler to search with the GenBank accession number or LOCUS name. Each sequence deposited in GenBank is associated with one LOCUS name and at least one accession number. For each LOCUS name or accession number, you will generally get just one sequence. Thus, you know roughly how many sequences you will get back from GenBank. To search GenBank by using accession numbers or LOCUS names or a combination of the two, just click the top option button and type in the accession numbers and/or LOCUS names, separated by a comma.

There are two output formats that you can choose. GenBank sequences can be delivered to your computer in either GenBank format or FASTA format. The FASTA format is one of the simplest sequence formats and sequences in this format can be delivered to your computer in a shorter time compared to sequences in GenBank format. However, sequences in FASTA format carry little information specific to the sequences, which severely restricts sequence analysis. For example, the coding region of the EF-1α gene is made of several exons interspersed in long stretches of introns. When you retrieve the sequences in FASTA format, you get a whole sequence with no specification on where each exon begins and ends. Consequently you will not be able to translate the nucleotide sequence into an amino acid sequence, and cannot use any codon-based or amino acid-based phylogenetic methods. Besides, because of the variation in intron lengths, you will have trouble aligning the sequences. Only when you know that you want to work on the entire sequences should you choose the FASTA format.

In contrast to the FASTA format, sequences in the GenBank format contain detailed annotation about the sequences in the FEATURES table, which is briefly explained in the previous chapter. DAMBE takes advantage of this information to splice out and join the coding sequences of the gene. The GenBank format is selected in this exercise (fig. 1).

You may also specify whether you wish to get nucleotide sequences or amino acid sequences. The former will search through the GenBank databases of nucleotide sequences, and the latter will search the databases of amino acid sequences.

Click the **Retrieve** button and the search will begin. Some sequences in the GenBank could be as long as several megabytes, and consequently could take a long time before the sequences were fully delivered to your computer. Once the target sequences have been retrieved, a standard **file/save** dialog will appear for you to save the retrieved sequences. Save the sequnces to a file. You will be presented with another dialog box (fig. 2). Because we are interested only in coding sequences, just click the **CDS** button and click **Proceed**.

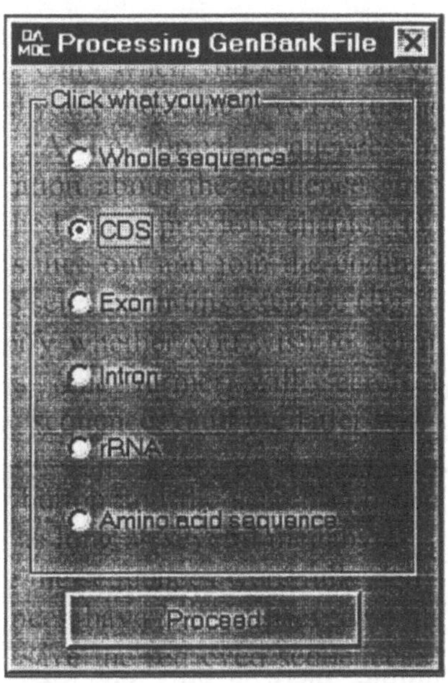

Figure 2. Specifying types of sequences to read into DAMBE

Another interactive dialog (fig. 3) is shown. There are five list boxes, with first column showing the LOCUS name of each GenBank sequence, the second showing the length of each sequence, and the third being taken from the DEFINITION entry of the GenBank sequence. The fourth and the fifth list boxes are currently empty. What you wish to get out of the GenBank file is specified under **Splice**, which is CDS for this operation. Note that the search specification with the word "cytochrome" is not very specific and the retrieved sequences could be either cytochrome b sequences or cytochrome oxidase subunit I, II, or III. Suppose we are really just interested in the coding sequences (CDS) of the cytochrome b gene.

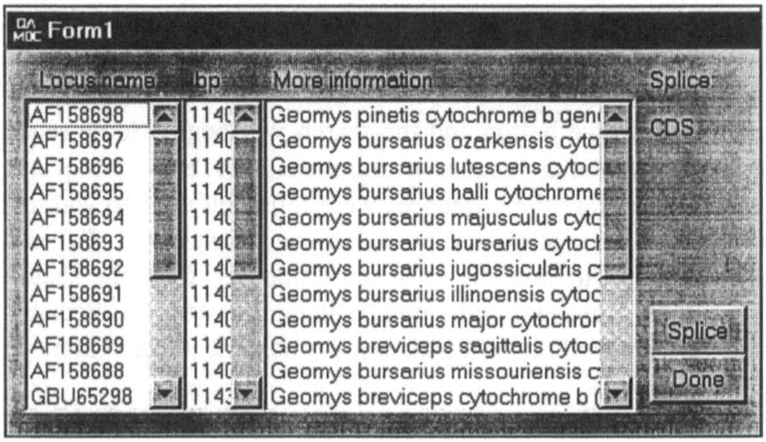

Figure 3. Reading GenBank file by taking advantage of the information contained in the FEATURES table. The dialog box is too wide to be shown in its entirety and only the left upper part is shown.

There are also some hidden boxes. For example, some sequences were deposited as complementary strand, and the GenBank file will so specify in the FEATURES table. DAMBE will take this information and automatically get the correct opposite strand, i.e, the actually transcribed RNA sequence. In this case, a text box with the word COMPLEMENT will be displayed. Because our sequences are not the complementary sequences, this text box is hidden.

Now click the first LOCUS name, i.e., AF158698. The dialog will change to display sequence-specific information for the LOCUS AF158698 (Fig. 4). The fourth list box displays the name of the target CDS sequence in AF158698. In our example, there is only one CDS named cytochrome b made of a continuous stretch of DNA. If the gene is made of several segments separated by introns, then the last listbox will display the numeric specification of the beginning and ending of each segment. The text box in the lower panel displays the complete sequence (Fig. 4).

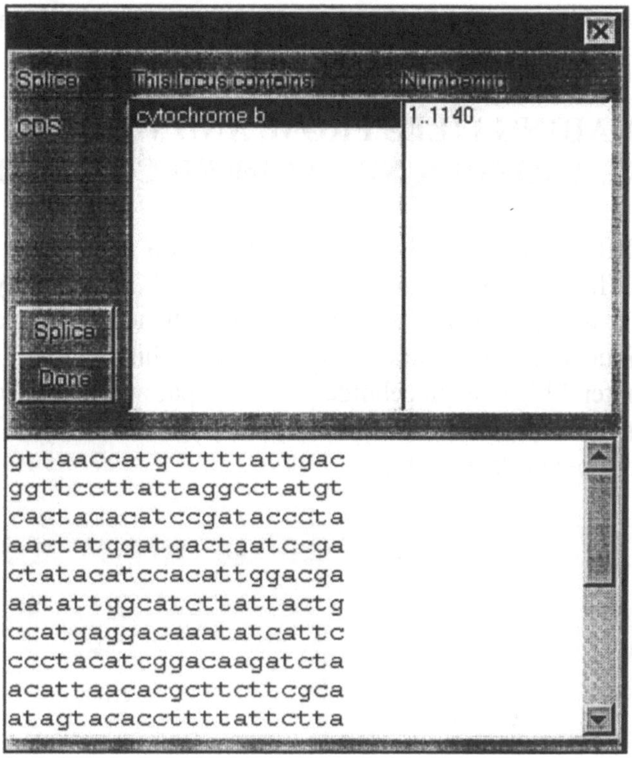

Figure 4. Dialog box displaying specifications for the coding sequence. Only the right half of the dialog box is shown because the dialog box is too wide.

Click the **Splice** button and the sequence will be read into DAMBE. When there are several dispersed exons, the **Splice** button will cut the exons out and join them together, so as to read in a complete coding sequence into DAMBE. Repeat this process for the rest of the LOCUSes. Click the **Done** button, and you will be prompted to confirm the type of sequences, which we have encountered several times already. Just click the option button **Protein-coding Nuc. Seq.** and then choose **Mammalian mitochondrial** as the genetic code.

Click the **Done** button, and the sequences are read into DAMBE, ready for you to apply any data analysis to them. For example, you may click **Phylogenetics|Distance methods|Nucleotide sequences**, and then click the **Done** button in the ensuing dialog box. A phylogenetic tree will be displayed, showing you the phylogenetic relationships among the DNA sequences that you have just retrieved.

We know that GenBank sequences can be retrieved in many ways, even through a web page. What is convenient in DAMBE is the it will immediate process the sequences and give you a set of sequences ready to run.

3. READING FILES FROM, AND WRITING FILES TO, ANOTHER NETWORKED COMPUTER

Start DAMBE and click **File|Reading data from networked computer**. A dialog box (fig. 5) appears for you to type in the IP address of your server, your login ID and password. Your password will not be displayed in the **Password** input field. Any character you type into this field will be replaced by the character "*" as a placeholder. If your password is "GoDAMBE", then the password input field will only display "*******". This is for protecting your network security.

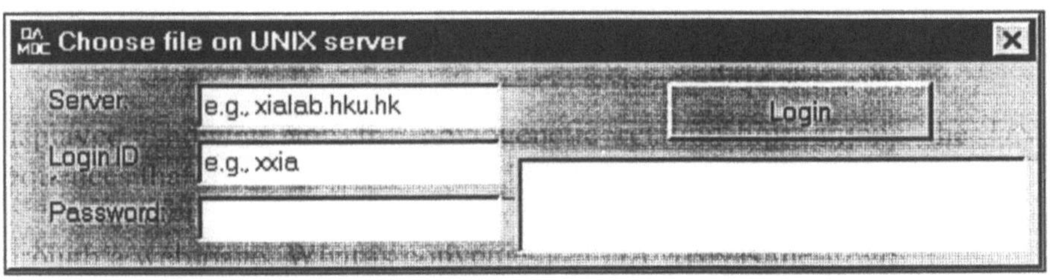

Figure 5. Dialog box for logging in to your networked computer

Click the **Login** button and DAMBE will go through the regular checking of your networked computer and the validity of your user ID and password. The text window below the **Login** button will display the interaction between your computer and the networked computer. Once the connection is established, which is generally guaranteed if you have entered the server IP, login ID and password correctly, DAMBE will present a dialog box (fig. 6) displaying the directory of your server, as well as other input fields relevant for reading files from your networked computer.

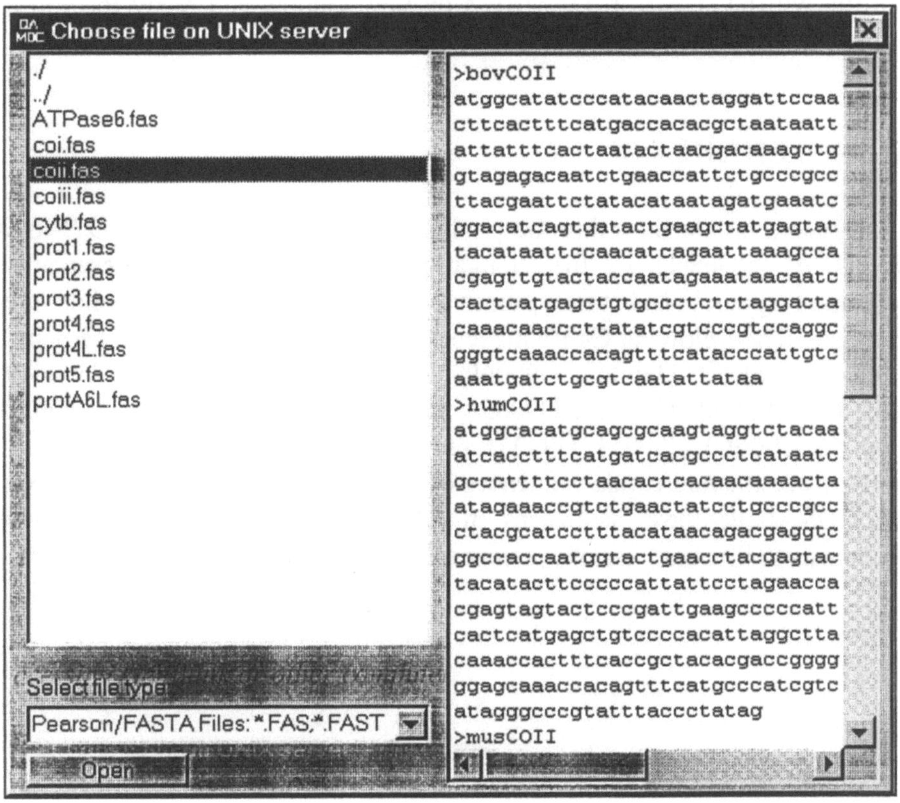

Figure 6. Reading sequence files from your networked computer

The dialog box contains four components. On the top-left is the directory structure, where an item followed by "/" is a subdirectory, otherwise it is a file. The first line "./" means the current directory, and the second line "../", when clicked, will bring you one directory up. Clicking any directory will bring you into that directory, clicking any file will have the file content displayed in the display window on the right side. The content of the file **coii.fas** was displayed.

What files are displayed is controlled by the **Select file type** dropdown box. The default is to display files with file extension **.FAS** or **.FASTA** (either in upper case or lower case). Clicking the downward arrow will show all other file types. For example, if you choose **GCG multiple sequence format (.MSF)**, then only files with file extension **.MSF** will be displayed in the directory listbox.

Click the **Open** button located at the lower left conner of the dialog box, and DAMBE will read in the highlighted sequence file, e.g., **coii.fas** in fig. 6, just as you would read in a file from your local hard drive. The content of the file will be shown in DAMBE's display window and ready for you to apply analytical methods to it.

4. EXERCISE

A recent article on arthropod phylogeny (Regier and Shultz 1997) used the DNA sequences of the elongation factor 1-α from representative arthropod species and their purative sister taxa such as annelids and molluscs. The following GenBank accession numbers are listed in the paper: U90045, U90052, U90047, U90048, U90055, U90053, U90057, U90049, X03349, U90058, U90054, U90059, U90062, U90063, U90046, U90050, U90056, U90060, U90051, U90061 and U90064.

Retrieve the nucleotide sequences of the EF-1α sequences and splice out the CDS. Save these sequences in FASTA format. We will use these sequences to illustrate various factors affecting phylogenetic reconstruction.

Chapter 5

Pair-wise and Multiple Sequence Alignment

1. INTRODUCTION

This chapter will only cover the most basic concepts of pair-wise and multiple sequence alignment so that you know what to do when using DAMBE to align your nucleotide or amino acid sequences. DAMBE uses codes in ClustalW (Thompson et al. 1994) for pair-wise and multiple sequence alignment. However, the implementation is rudimentary. One advantage of DAMBE over ClustalW is when you wish to align protein-coding nucleotide sequences against aligned amino acid sequences so that you will not have the annoying frameshifting indels introduced as alignment artefact.

1.1 The dot-matrix approach

The dot-matrix method is for quickly aligning sequences that are very similar. For illustration purpose, let's start from the simplest case, with two identical sequences:

Seq1: ATTCCGGTACGT
Seq2: ATTCCGGTACGT

Write the two sequences to be aligned as row and column headings in the matrix (or grid) below. Look from left to right. If a column heading matches the row heading, then put a dot at the intersecting cell. This results in the "dot-matrix" shown below.

	A	T	T	C	C	G	G	T	A	C	G	T
A	•								•			
T		•	•					•				•
T		•	•					•				•
C				•	•					•		
C				•	•					•		
G						•	•				•	
G						•	•				•	
T		•	•					•				•
A	•								•			
C				•	•					•		
G						•	•				•	
T		•	•					•				•

Note that there is a dot in all 12 diagonal cells, i.e., the dots can be connected by a straight line. This means that there is no gap (indels) needed for aligning the two sequences.

Now suppose the two sequences are slightly different:

Seq1: ATT**C**CGGT**G**CG**T**
Seq2: ATT**G**CGGT**A**CG**A**

Again write the two sequences to be aligned as row and column headings in the matrix (or grid) below and put a dot in the cell in which a column heading matches the row heading. Now we have nine dots in the 12 diagonal cells. When most of the diagonal cells are filled with dots that can be connected by a straight line, there is no need for inserting gaps (indels) although some bases do not match.

	A	T	T	C	C	G	G	T	G	C	G	T
A	•											
T		•	•					•				•
T		•	•					•				•
G						•	•		•		•	
C				•	•					•		
G						•	•		•		•	
G						•	•		•		•	
T		•	•					•				•
A	•											
C				•	•					•		
G						•	•		•		•	
A	•											

Now suppose we have two messy sequences like the following:

Seq1: ATTCCGGTACGT
Seq2: ATTCCAAAGGTACGT

Following the previous protocol of putting down dots in the grid, you will arrive at the following dot-matrix:

	A	T	T	C	C	G	G	T	A	C	G	T
A	•								•			
T		•	•					•				•
T		•	•					•				•
C				•	•					•		
C				•	•					•		
A	•								•			
A	•								•			
A	•								•			
G						•	•				•	
G						•	•				•	
T		•	•					•				•
A	•								•			
C				•	•					•		
G						•	•				•	
T		•	•					•				•

Note that there are consecutive dots in the first five diagonal cells, and there are consecutive dots in the last seven diagonal cells. However, we cannot connect the 12 dots with a straight line. This means that there must be indels somewhere. Note that we can connect the first five dots, shift down three cells, and then connect the last seven dots. This shifting down of three cells, after the fifth base, means the insertion of three gaps in Seq1 after the fifth base. So we have:

```
Seq1: ATTCC---GGTACGT
Seq2: ATTCCAAAGGTACGT
```

Note that if we need to shift **horizontally** three cells, rather than shifting **down** three cells, then it would mean an insertion of three gaps in Seq2.

Let us now deal with a more complicated case. Suppose we have the follow sequences that I have taken from Li (1997):

```
Seq1: ATGCGTCGTT
Seq2: ATCCGCGTC
```

Now work out the dot matrix and decide what alignment you should choose. You will find two alternative alignments, shown below, that are better than others:

Alignment 1:

```
Seql: AT--GCGTCGTT
      ||  |||||
Seq2: ATCCGCGTC---
```

Alignment 2:

```
Seql: ATGCGTCGTT
      || || |||
Seq2: ATCCG-CGTC
```

Both alignments have seven matched pairs, but the first alignment has more gaps (indels). Because the indels are generally considered rare, we tend to take the second alignment as more likely. This brings in the concept of gap penalty that will be dealt with in the next section.

Note that in all dot-matrix illustrations above, each nucleotide occupies a cell. In practice, one can often divide long sequences into equal-length segments and each segment can then occupy a cell. The dots will then represent matched segments. If the segment contains n nucleotides or amino acids, then each segment is one n-tuple. The default value for n in ClustalW (and also in DAMBE) is three for nucleotide sequences and one for amino acid sequences.

1.2 Similarity or distance method

Let's first define a similarity index (S) that measures the similarity between the two sequences to be aligned:

$$S = Max(n_m - \sum_{i=1}^{k_{Max}} w_k n_k) \tag{5.1}$$

where n_m is the number of matched pairs, w_k the penalty for a gap of k nucleotides, n_k the number of gaps with length k, and k_{Max} the maximum gap length allowed. The gap penalty w_k is expressed as an increasing function of k, with its simplest form being $w_k = a + b \bullet k$. The similarity method of sequence alignment is a protocol of finding an alignment with the maximum S. It is clear that if we use different values for a and b, we may get different optimal alignments because S is a function of a and b. This is why an alignment program will typically prompt the user for input on the gap penalty.

The distance method of sequence alignment is similar. First we define a distance index (D) that measures the difference between the two sequences to be aligned:

$$D = Min(n_d + \sum_{i=1}^{k_{Max}} w_k n_k) \tag{5.2}$$

The distance method is a protocol for finding the alignment with the smallest D. One of the advantages of using the distance method is that, in multiple sequence alignment, the matrix of pair-wise D values can be used to find an approximate phylogenetic tree by using any one of the distance methods in phylogenetic reconstruction. For example, CLUSTALW uses neighbor-joining method to find the phylogenetic tree before proceeding to multiple sequence alignment.

2. SEQUENCE ALIGNMENT USING DAMBE

Sequence alignment is slow, especially with long nucleotide sequences. For this reason, you should start with short and few sequences just to get a feel of how time-consuming DAMBE is when doing multiple alignment. Aligning amino acid sequences is faster than aligning nucleotide sequences.

2.1 Align nucleotide or amino acid sequences

Start DAMBE and open a file containing unaligned sequences. You will be told that the sequences are unaligned and whether you wish to align the sequences. Click **Yes**, and a dialog box (fig. 1) will appear for you to specify options. For a first try, it is better just use all default options. Note that the dialog box in fig. 1 is for nucleotide sequences. If the input file contains amino acid sequences, then the **DNA weight matrix** will be replaced by **Protein weight matrix**. Click the **Go!** button and wait for a while. You will then be asked if you wish to save the aligned sequences. Choose **Yes** and save the file. The aligned sequences will then replace the original sequences in DAMBE's buffer.

Procedures for aligning amino acid sequences are the same as those for aligning nucleotide sequences. If you have read in a set of unaligned sequences without first aligning them, you can do the alignment by clicking **Sequences|Align sequences using ClustalW**.

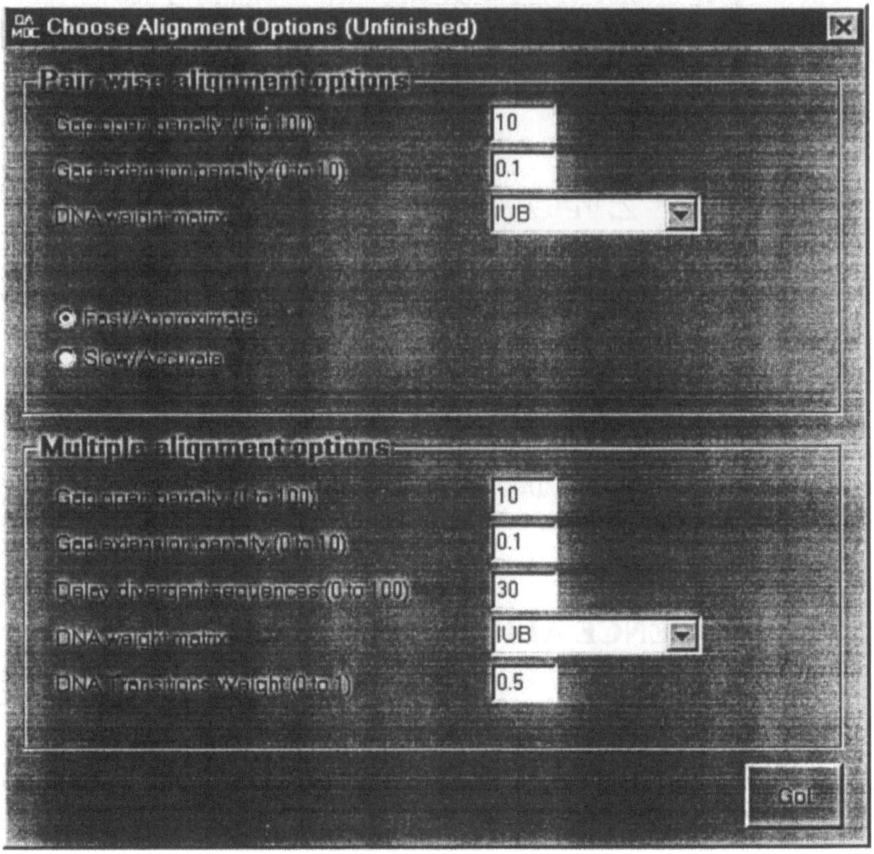

Figure 1. Aligning nucleotide sequences

2.2 Align nucleotide sequences against amino acid sequences

One frustrating experience I have often had with aligning protein-coding nucleotide sequences is the introduction of many frameshift indels in the aligned sequences, even if the protein genes are known to be all functional and do not have these frameshifting indels. In other words, the introduced frameshifting indels in the aligned sequences are alignment artefacts, and the correctly aligned sequences should have complete codons, not one or two nucleotides, inserted or deleted.

One way to avoid the above alignment problem is to align the protein-coding nucleotide sequences against amino acid sequences. This obviously requires amino acid sequences which can be obtained in two ways. First, if you have nucleotide sequences of good quality, then you can translate the sequences into amino acids. Second, if you are working on nucleotide sequences deposited in GenBank, then typically you will find the corresponding translated amino acid sequences. DAMBE can read both the nucleotide sequence and the corresponding amino acid sequence in a GenBank sequence.

Here I illustrate the use of this special feature by assuming that you already have a file containing unaligned protein-coding nucleotide sequences, say **unaligned.fas**, in your hard disk.

Start DAMBE, and open the **unaligned.fas** file. When asked whether to align the sequences, click **No**. The unaligned sequences will then be read into DAMBE's buffer. Now click **Sequences|Work on Amino Acid Sequences** to translate the protein-coding nucleotide sequences into amino acid sequences. If the translation results in a number of termination codons embedded in the sequences (represented by "*"), then either your nucleotide sequences are of poor quality or they might be from pseudogenes. In either case you should give up aligning your nucleotide sequences against these junky amino acid sequences.

If the translation looks good, then click **Sequence|Align sequences with Clustal** to align the translated amino acid sequences. Once this is done, you have a set of aligned amino acid sequences in the DAMBE buffer for you to align your nucleotide sequences against.

Click **Sequence|Align nuc. seq. against aligned aa seq**. A standard file **Open/Save** dialog box will appear. Choose the **unaligned.fas** file again, which contains the unaligned nucleotide sequences. DAMBE will align the nucleotide sequences against the aligned amino acid sequences in the buffer. This procedure ensures that no frameshifting indels are introduced as an alignment artefact.

If your sequences were retrieved from GenBank, then most protein-coding genes will already have translated amino acid sequences included in the FEATURES table of GenBank files. You can use DAMBE to first read in all amino acid sequences, align these amino acid sequences, and then ask DAMBE to splice out the corresponding CDS, and align the CDS sequences against aligned amino acid sequences in DAMBE buffer.

Chapter 6

Factors Affecting Nucleotide And Di-Nucleotide Frequencies

1. INTRODUCTION

You have now familiarized yourself with molecular data in various forms. The rest of the book will focus on the analysis of molecular sequences. Molecular sequences are information-rich and there are many different ways of extracting useful information out of them. For evolutionary biologists who wish to obtain historical information from DNA, one of the main objectives of studying molecular sequences is to understand the dynamic nature of the sequences, i.e., how the molecular sequences change over geological time. The change of molecular sequences over time is typically characterized by a substitution model, and how well the model fits the observed substitution pattern reflects how well we have understood the dynamic nature of molecular sequences.

A model, be it a substitution model or any other model, has at least one parameters. For example, a linear equation, $Y = \alpha + \beta X$, has two parameters, i.e., the intercept α and the slope β. A substitution model has two categories of parameters, the frequency parameters and the rate ratio parameters. This chapter and the next two chapters will deal specifically with the frequency parameters, and the rate ratio parameters will be introduced in later chapters.

1.1 The frequency parameters

There are four nucleotides, 20 amino acids, and about 60 sense codons. The frequency parameters refer to the frequencies of these nucleotides, amino acids, or codons in nucleotide, amino acid or codon sequences, respectively. If you have encountered a few nucleotide-based substitution models, such as the JC69 (Jukes and Cantor 1969) and the K80 (Kimura 1980) models, you might have already realized that these models assume equal nucleotide frequencies at equilibrium. Such substitution models have no frequency parameters to be estimated because they are all assumed to be 0.25, which you will soon find out to be a very unrealistic assumption. The simplest substitution models that allow frequency parameters to be different from each other are the F81 (Felsenstein 1981) and the TN84 (Tajima and Nei 1984) models. Because frequency parameters have to sum up to one, there are only 3 (= 4 - 1) frequency parameters to be estimated for nucleotide sequences and 19 frequency parameters for amino acid sequences. For codon-based models, if we ignore termination codons, then the number of frequency parameters is the number of sense codons minus one.

If you have already explored a substantial part of the landscape of molecular biology and evolution, you might have come across terms such as stationarity. If the stochastic process governing the evolution of nucleotide, amino acid or codon sequences does not alter the nucleotide, amino acid or codon frequencies, respectively, then the stochastic process is stationary. A stationary process implies that the probability of a particular substitution, e.g., an A→G transition, remains the same during the evolutionary history. All substitution models implemented in existing computer programs based on the maximum likelihood method share the stationarity assumption. If the assumption is violated, then the results from phylogenetic analyses using the maximum likelihood method should be interpreted with caution.

1.2 Factors that might change the frequency parameters

Different genomes often differ much in nucleotide and dinucleotide frequencies, and some of these differences can be interpreted in light of molecular adaptation. One such example concerns the evolution of GC content in prokaryotes. Thermophilic bacteria tend to have high GC content, which can be interpreted as an adaptation against DNA denaturation under high temperature because G/C pairs have three hydrogen bonds whereas A/T pairs have only two. Another interpretation is that GC-rich codons tend to code for thermally stable amino acids and GC-poor codons tend to code for thermally unstable amino acids (Argos et al. 1979).

An alternative interpretation of the association of high GC-content with high ambient temperature is that ancient organisms are GC-rich, and these GC-rich bacteria have maintained their GC-richness because of their evolutionary inertia. I do not think this explanation plausible. Those thermophilic bacteria are not mutation free. In all likelihood they should have higher mutation rate due to the high temperature. Random mutations tend to favour AT, not GC. In short, these organisms are not expected to maintain their GC-richness without involvement of selection.

Another evolutionary hypothesis that is related to dinucleotide frequencies in the genome concerns T-T dimers produced by UV radiation (Singer and Ames 1970). It is hypothesized that organisms exposed to sunlight should have a lower frequency of genomic TT dinucleotides to avoid the deleterious effect of T-T dimers than those that do not expose to sunlight (e.g., intestinal bacteria such as E. coli). This hypothesis can be tested by examining dinucleotide frequencies between organisms exposed to UV light and those that do not.

Vertebrate genomes tend to show a dramatic deficiency in CpG dinucleotides (CpG is shorthand for 5'-CG-3'), which is a likely consequence of heavy DNA methylation in vertebrate genomes. DNA methylation is a ubiquitous biochemical process observed in both prokaryotes (Noyer-Weidner and Trautner 1993) and eukaryotes (Antequera and Bird 1993). In vertebrates, DNA methylation mainly involves the methylation of C in the CpG dinucleotide, which greatly elevates the mutation rate of C to T through spontaneous deamination of the resultant 5-methylcytosine (Barker et al. 1984; Cooper and Krawczak 1989; Cooper and Krawczak 1990; Cooper and Schmidtke 1984; Cooper and Youssoufian 1988; Ehrlich 1986; Ehrlich et al. 1990; Rideout et al. 1990; Schaaper et al. 1986; Sved and Bird 1990; Wiebauer et al. 1993). Thus, the origin of methylation must have resulted in a nonstationary process in which the probability of C→T transitions is substantially increased. Because a C→T transition in one strand will lead to a G→A transition on the opposite strand, we expect DNA methylation to reduce GC content and increase AT content.

1.3 Frequency parameters and phylogenetic analyses

It is also important to appreciate the variation in nucleotide frequencies among different organisms when you work on phylogenetic reconstruction, for three reasons. First, some measures of genetic distances assume equal nucleotide frequencies of ¼, e.g., those models characterized by a symmetrical substitution rate matrix, such as that for Jukes and Cantor's (1969) one-parameter model and Kimura's (1980) two-parameter model. If nucleotide frequencies differ much from ¼, then these distances would not be appropriate.

Second, when your OTUs have diverged for a long time, then multiple substitutions would occur at the same nucleotide site, leading to what is know as substitution saturation. Obviously, when sequences have reached full substitution saturation, then their similarity will depend entirely on similarity in nucleotide frequencies. Because the similarity in nucleotide frequencies has little to do with phylogenetic relationship, the phylogenetic trees you produce would be misleading. When your sequences have diverged for a long time, then it is very important to examine the nucleotide frequencies to see if the resulting phylogenetic relationship exhibits any dependence on the nucleotide frequencies. I will illustrate this issue with a case study after this chapter.

Third, even if your OTUs have not experienced substitution saturation, they may differ much in nucleotide frequencies. Most computer programs for phylogenetic reconstruction assume a stationary Markov process of nucleotide substitution with the nucleotide frequencies remaining unchanged during sequence divergence. When different OTUs have very different nucleotide frequencies, this assumption obviously is violated and the result of phylogenetic analysis based on these programs will be misleading. It is important to at least have a look at the difference of nucleotide frequencies of your OTUs before subjecting them to computer programs for phylogenetic reconstruction. A few substitution models have relaxed this assumption of stationarity. For example, the models underlying Lake's (1994) paralinear distance and the LogDet distance by Lockhart et al. (1994) which is similar to Lake's distance are presumably applicable to the situation when nucleotide frequencies have changed during the divergence of different lineages. They should be used in situations where nucleotide frequencies differ much among OTUs. Lake's paralinear distance is implemented in DAMBE for nucleotide sequences. We will learn later how to use the distance, as well as many other distances implemented in DAMBE, to carry out phylogenetic analysis.

Please don't skip this chapter just because you think that you'll never need to do something as simple as counting nucleotide and dinucleotide frequencies. This chapter has two objectives. The first is to let you have a better appreciation of the questions of molecular evolution and phylogenetics mentioned above. The second is to get you familiar with DAMBE's user interface. You will feel more comfortable with DAMBE after going through this chapter.

2. COUNTING NUCLEOTIDE AND DINUCLEOTIDE FREQUENCIES WITH DAMBE

Start DAMBE, and open a file containing nucleotide sequences (if you have not done so already). If you do not have your own sequences, then just open the **virus.fas** file that came with DAMBE. The file is in the same directory as **dambe.exe** and contains protein-coding nucleotide sequences of a hemaglutinin gene from Influenza A viruses infecting mammalian species. The viral gene shares the same universal genetic code as the nuclear genome of their mammalian host.

Click the **Seq. Analysis** menu, and then click the Nucleotide Frequency menu item. A dialog box appears (fig. 1). Such a dialog box (or slight variation of it) will also appear when you click other menu items under the **Seq. Analysis** menu. So take this opportunity to become familiar with it. There are two lists in the dialog box. The one on the left shows the sequences that are available for selection. The one on the right displays sequences selected for computing nucleotide and di-nucleotide frequencies. At this moment, the list on the right is empty.

- To select a single sequence, just click to highlight it, and then click the → button to move it to the right.
- To select neighboring sequences, click the first of the neighboring sequences to highlight it and then, holding down the shift key, click the last of the neighboring sequences. All the neighboring sequences will then be highlighted. Click the → button to move them to the right.
- To selection disjoint sequences, click each sequence while holding down the Ctrl key, and then click the → button to move the highlighted sequences to the right.

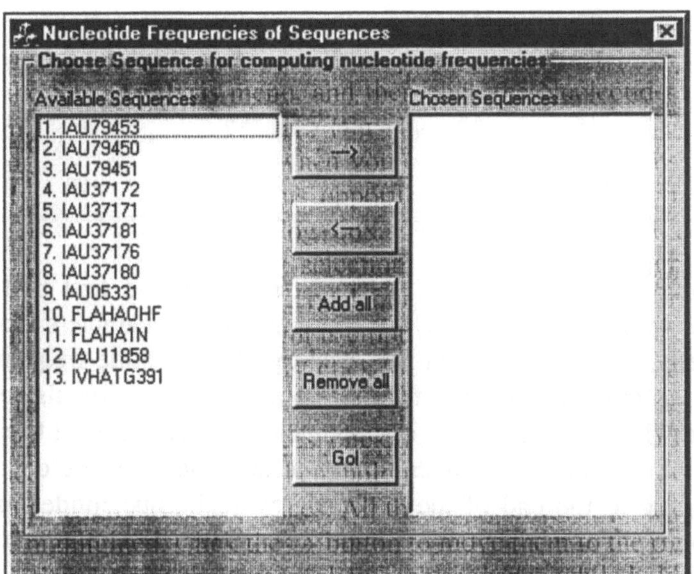

Figure 1. Dialog box for selecting molecular sequences for analysis.

Once you have finished your selection, click the **Done** button. After a few seconds, the standard **file/open** dialog box appears. Type in a file name for saving the result, or simply use the default. Then click the **Save** button. The file is then saved in text format, and also displayed in the display window. A part of a sample output (for one sequence) is shown below:

	A	C	G	U	Other	Sum(ACGU)
FLAHAOHF						
Freq	339	211	209	210	0	969
Prop.	.35	.22	.22	.22		1

	A	C	G	U	Sum
Obs. A	117	75	72	75	339
Exp.	119	74	73	74	
Obs. C	88	52	26	44	210
Exp.	74	46	45	46	
Obs. G	79	32	53	45	209
Exp.	73	46	45	45	
Obs. U	55	52	57	46	210
Exp.	74	46	45	46	
Subtotal	339	211	208	210	968

The output is of two parts for each sequence, the first part lists the nucleotide frequencies, with "Other" stands for all characters that are not "acgtu", e.g., "-?.". The second part lists the di-nucleotide frequencies, and the expected frequencies when there is no association or repulsion between nucleotides (i.e., the probability of two nucleotides sitting next to each other depends entirely on their frequencies). The di-nucleotides are counted from the beginning to the end of the sequence, with the nucleotides on the left column being the first, and those on the top row being the second, of the dinucleotide.

From the first part of the output, we note that A is used more frequently than other nucleotides. Should you choose the JC69 or the K80 model for data analysis involving this group of sequences?

The high frequency of nucleotide A might be caused by the preponderance of Asn and Lys (which are coded by AAN codons, where N stands for any of the four nucleotides) in the protein-coding gene, or it might be caused by codon usage bias favoring A-ending codons for more efficient transcription (Xia 1996). Which of these two possible scenarios is likely true can be revealed by going through the chapter on computing codon frequencies.

We also note that there is a deficiency of CG dinucleotides (fewer then expected), and a surplus of TG dinucleotides (more than expected). This pattern is readily explainable by invoking DNA methylation in mammalian species. The gene, labelled FLAHAOHF, is a hemaglutinin gene from an Influenza A virus infecting mammalian species, all of which seem to show high levels of DNA methylation. DNA methylation is a ubiquitous biochemical process observed in both prokaryotes (Noyer-Weidner and Trautner 1993) and eukaryotes (Antequera and Bird 1993). In vertebrates, DNA methylation mainly involves the methylation of C in the CpG dinucleotide, which greatly elevates the mutation rate of C to T through spontaneous deamination of the resultant 5-methylcytosine (Cooper and Krawczak 1989; Cooper and Krawczak 1990; Rideout et al. 1990; Sved and Bird 1990; Wiebauer et al. 1993). This biochemical process accounts for the deficiency of CG dinucleotides and a surplus of TG dinucleotides in the gene. You see that even a very simple procedure can lead us to some biological insights.

One of the assumptions of most substitution models implemented in computer programs states that substitutions occur independently in different sites. This assumption is no longer tenable given the effect of DNA methylation, because the substitution of C by T occurs at a much faster rate with a downstream neighboring G than with the G being replaced by other nucleotides.

Did you notice that the TA dinucleotide is also rare? Can you think of some reasons to explain its rarity? In protein-coding genes, a TA dinucleotide can occur only in three situations, in TAN codons, NTA codons, and NNT ANN double codons, where N stands for any of the four dinucleotides. We know that TAR are termination codons that can occur only once in a protein-coding sequence. This might contribute to the rarity of TA dinucleotides. Now you see that to understand nucleotide and dinucleotide frequencies, it is often necessary to know codon usage bias. The chapter after the next will introduce you to the interplay of mutation and selection operating on codons.

Chapter 7

Case Study 1

Elongation Factor-1α And The Arthropod Phylogeny

1. INTRODUCTION

This case study illustrates the consequence of ignoring differences in nucleotide frequencies among species in phylogenetic studies. To fully understand this case study, you would need to use phylogenetic methods in DAMBE that we have not yet covered. However, sometimes it is beneficial to do something ahead of time.

Elongation factor-1α (EF-1α) is one of the most abundant proteins in eukaryotes (Lenstra et al. 1986) and catalyzes the GTP dependent bindings of charged tRNAs to the ribosomal acceptor site (Graessmann et al. 1992). Because of its fundamental importance for cell metabolism in eukaryotic cells, the genes coding for the protein are evolutionarily conservative (Walldorf and Hovemann 1990), and consequently have been used frequently in resolving deep-branching phylogenies such as the arthropod phylogeny (Regier and Shultz 1997).

Arthropods are conventionally classified into three major groups: Atelocerata (Hexapoda + Myriapoda), Chelicerata, and Crustacea (Branchiopoda + Malacostraca). The phylogenetic relationship among these three major groups have been controversial (Fryer 1998), and the controversy is further complicated by a recent claim that Atelocerata and Crustacea are both polyphyletic (Regier and Shultz 1997). Specifically, Hexapoda and Myriapoda do not form a monophyletic group, neither do Branchiopoda and Malacostraca.

Arthropods have been around for a long time, and the sequences used in (Regier and Shultz 1997) have diverged so much that substitution saturation is substantial, especially at the third codon positions. Recall that sequence similarity becomes highly dependent on similarity in nucleotide frequencies in sequences having experienced substitution saturation. We will use their data set to illustrate the problem of heterogeneity in nucleotide frequencies among species in phylogenetics.

2. OBTAIN DATA FROM GENBANK

Molecular sequences can be retrieved from GenBank by using either GenBank accession number or LOCUS name, or by using keyword searching. The accession numbers for the EF-1α gene sequences from invertebrate species are U90045, U90052, U90047, U90048, U90055, U90053, U90057, U90049, X03349, U90058, U90054, U90059, U90062, U90063, U90046, U90050, U90056, U90060, U90051, U90061 and U90064. These are taken from Table 1 in Regier and Shultz (1997). Accession numbers and GenBank LOCUS names are unique identification tags for sequences deposited in GenBank. We will now retrieve these sequences from GenBank by using DAMBE's network function.

Start DAMBE and click **File|Read sequences from GenBank**. A dialog box (fig. 1) appears for you to specify options. The default is to search by using GenBank accession number or LOCUS name, which is what we want. The second option, shown in the lower left corner, concerns whether to get the retrieved file in the complicated GenBank format or the simple FASTA format. The default is GenBank format, which is also what we want. The third option allows the user to specify whether to search protein data banks or the nucleotide data banks. The latter is what we want. Enter the accession numbers listed in the previous paragraph into the text box, separated by a comma, as shown in fig. 1. Click the **Retrieve** button and the 21 sequences will be retrieved in GenBank format. When the retrieval is complete, which may take several minutes, you will be prompted to save the retrieved sequences. Just enter a file name and leave the **.GB** file type as default.

In the dialog box that comes next, click the **CDS** option and then click the **Proceed** button. CDS standing coding sequences, i.e., the nucleotide sequences that specify the amino acid sequence.

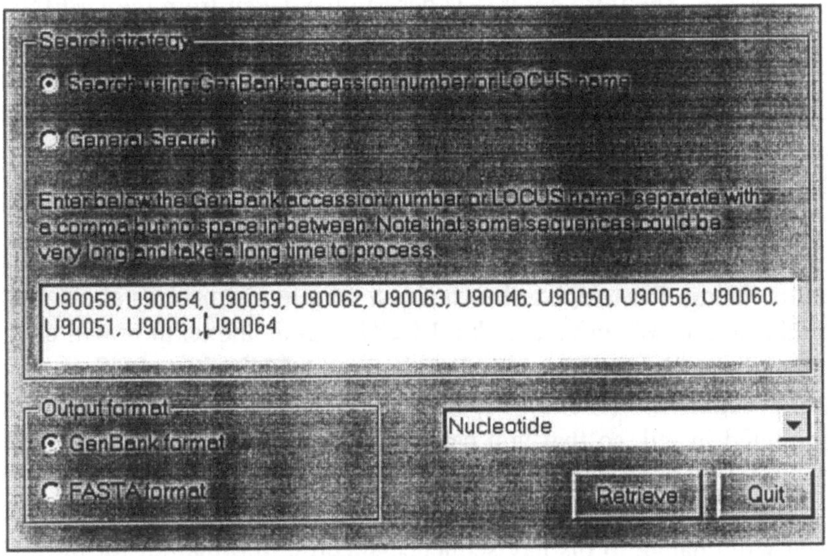

Figure 1. Retrieving sequences from GenBank. The top line, containing 9 accession numbers, is out of view.

You will now be staring at a seemingly rather complicated dialog box (fig. 2). Only part of the dialog box is shown. Let me explain each item briefly.

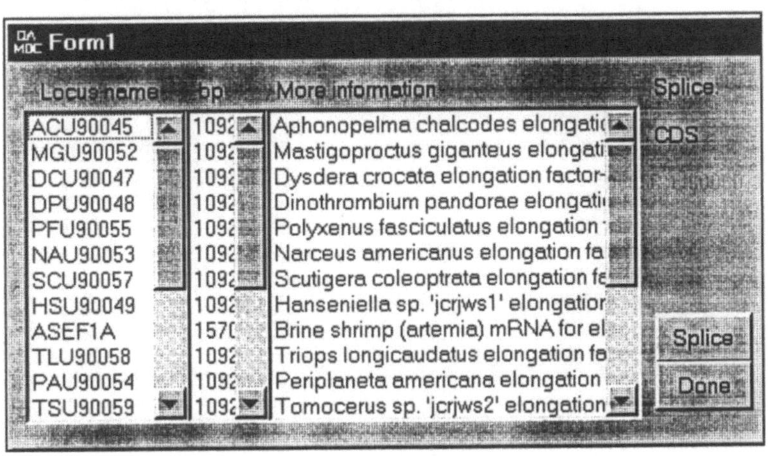

Figure 2. Processing sequences in GenBank format. Only left upper part of the dialog box is shown.

Each of the 21 sequences has a LOCUS name, which is shown in the first listbox (fig. 2). The second list box shows the length of the sequence for each LOCUS. The third column is descriptive, taken from the DEFINITION clause of the GenBank sequence. The right side of fig. 2, which is in fact in the middle of the dialog box if the whole dialog box is shown, shows specific segments that we are interested, i.e., CDS for coding sequences. There are also two listboxes (not shown) on the right of the dialog. These listboxes will show additional information when we click a LOCUS name in the first column.

There is also a textbox immediately below what is shown in fig. 2. For the time being, the textbox shown a few simple instructions about how you should proceed. When you click a LOCUS name, the textbox will change to display the sequence associated with the LOCUS. If a CDS is made of several segments (exons) separated by introns, these exons will be highlighted in red, so that you can visually see where each exon starts and ends.

Now click the first LOCUS in the left most listbox, i.e., ACU90045. The dialog will change to display sequence-specific information for the LOCUS ACU90045 (fig. 3). The fourth listbox (i.e., the left listbox in fig. 3) displays the name of the target CDS sequence in ACU90045. In our case, there is only one CDS named **elongation factor-1 alpha**. The "<" and ">" symbol in the fifth listbox tell us that the sequence does not contain the complete coding sequence of EF-1α, with "<" signalling a missing segment at the beginning and the ">" signalling the missing segment at the end.

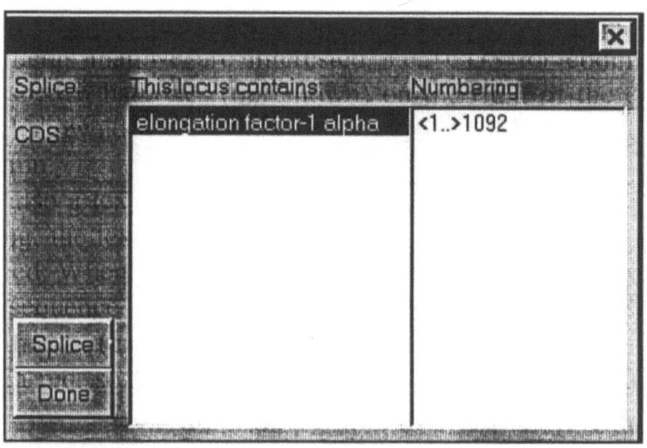

Figure 3. Processing sequences in GenBank format

Click the **Splice** button and the sequence is extracted from the GenBank file. If the CDS is made of several segments, then all these segments will be displayed in the last listbox, with each segment specified by a starting number and an ending number. When you click the **Splice** button, DAMBE will extract these segments and join them together to form a complete CDS specifying the amino acid sequence of the protein.

Once you have clicked the **Splice button,** the phrase "elongation factor-1 alpha in the fourth column will change to "Done". Click the next LOCUS name, and then click the **Splice** button again, until you have finished extracting all sequences. Click the **Done** button.

You will be prompted for sequence type. Click the option button for **Protein-coding nuc. sequences**. The dialog will expand to show the 12 genetic codes implemented in DAMBE. Click the **Universal** option and then the **Go!** button. You will be told that the sequences are not aligned, and asked if you wish to align the sequences using ClustalW. DAMBE incorporates most of the codes for pair-wise and multiple alignments in ClustalW. If you click **Yes**, then the nucleotide sequences will be aligned, but the alignment will be poor, with many frameshifting insertions and deletions (indels) resulting from artefacts introduced by ClustalW. DAMBE can align protein-coding nucleotide sequences against aligned amino acid sequences, which better then aligning nucleotide sequences directly. For this reason, Click **No** when prompted to align sequences.

You will be reminded that many functions in DAMBE assume that the sequences are aligned. Click **OK** to put it away. The extracted sequences for elongation factor-1 alpha (EF-1α) will be displayed. You should now save your sequences. To do so, click **File|Save as (convert sequence format)**, choose one of many output formats (e.g., FASTA) supported in DAMBE, type in a file name (e.g., **invert.fas**), and click the **Save** button.

3. ALIGN THE SEQUENCES

I mentioned that DAMBE can align nucleotide sequences against aligned amino acid sequences, but we do not yet have aligned amino acid sequences. So how should we proceed? Let me show you.

Click **Sequences|Work on amino acid sequences** to translate the nucleotide sequences into amino acid sequences. You may wonder how would DAMBE know where to begin translation. The answer is that DAMBE does not. What DAMBE does is simply do extra work by translating each nucleotide sequences into amino acid sequences at all three possible positions, i.e., starting from the first, the second, and the third nucleotide sites separately. A good protein-coding nucleotide sequence should have no embedded termination codon. DAMBE simply picks up the translation with the fewest termination codons.

Once the translation is done, you can now align the amino acid sequences by clicking **Sequence|Align sequences using ClustalW**. A dialog box appears (fig. 4). There are two sets of options that you can specify, one for pair-wise alignment and one for multiple alignment. If you do not know much about the values in the input fields, then just leave them as is for the time being and click **Go!**. The sequences will be aligned and displayed. Click **File|Save as (converting sequence format)** to save the aligned amino acid sequences in FASTA format.

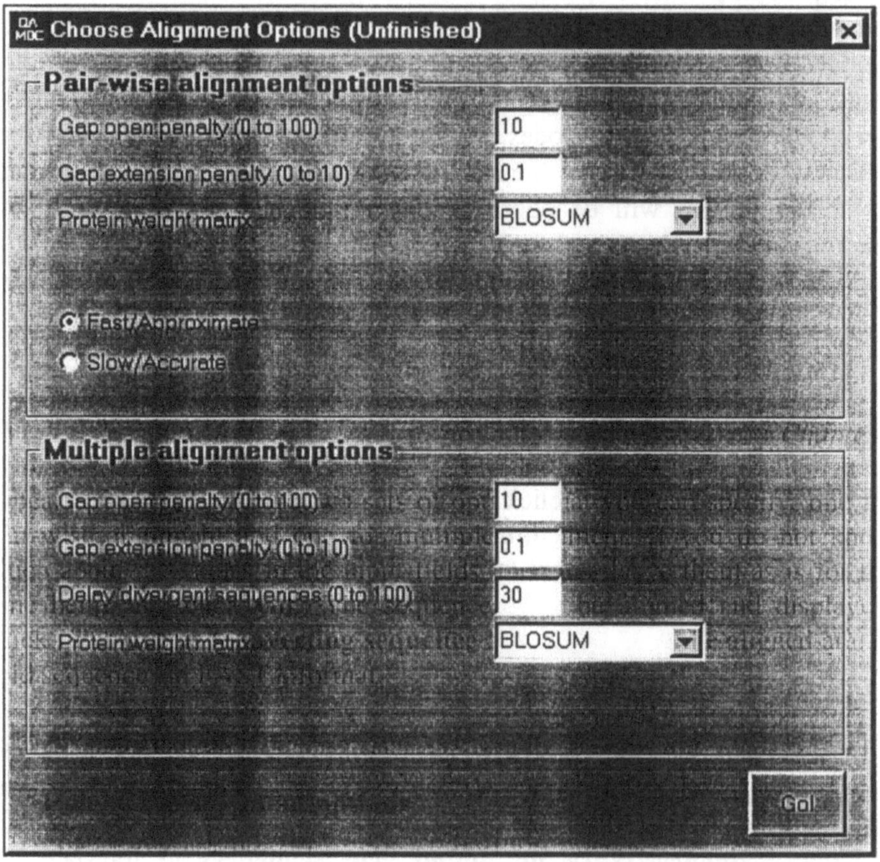

Figure 4. Dialog for aligning amino acid sequences

The protocol of multiple alignment is typically done as follows. First, all possible pair-wise alignment is done, and a dissimilarity score is generated for each pair of sequences. These dissimilarity scores are used as genetic distances to reconstruct a phylogenetic tree by using a distance method such as the neighbor-joining method. Multiple alignment is then carried out along the branches of the phylogenetic tree, starting from the more closely related taxa and progress to more distantly related taxa. The quality of the alignment therefore partially depends on the quality of the phylogenetic tree. After the multiple sequence alignment, DAMBE will ask you if you wish to look at the phylogenetic tree used for alignment. If you click **Yes**, then the tree is displayed (fig. 5).

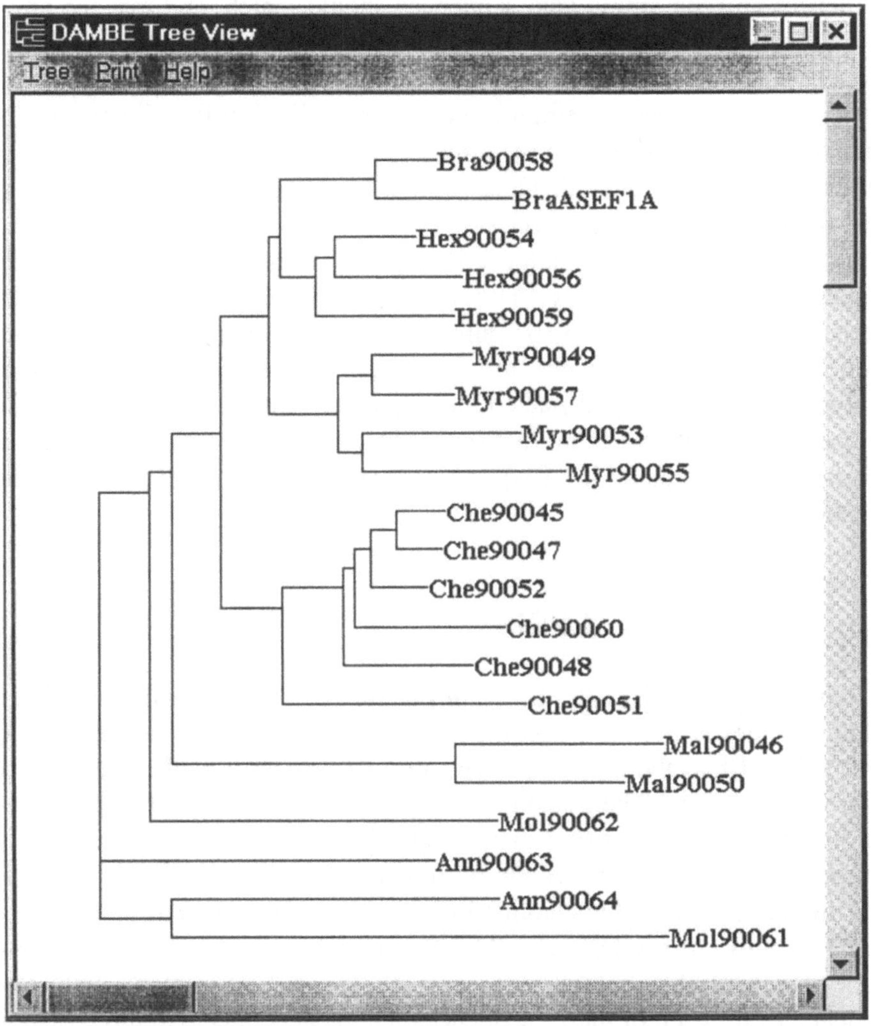

Figure 5. Phylogenetic tree used for multiple sequence alignment

The tree looks reasonable, except for the ancient divergence involving Annelids and Molluscs. The tree can be rerooted or printed in high quality, but we have little time to lose. So please just click **Tree|Exit** to quit the tree-display window.

We have not yet finished with sequence alignment. What we have just done is to obtain aligned amino acid sequences, which are now residing in

DAMBE's buffer. We still have not aligned the nucleotide sequences against the aligned amino acid sequences.

Click **Sequence|Align nuc sequences to aligned AA sequences in buffer**. You will be warned that the sequence names for the nucleotide sequences and the amino acid sequences should match, and that the amino acid sequences in DAMBE's buffer are really translated from the nucleotide sequences. If the amino acid sequences are for haemoglobin genes while nucleotide sequences are albumin genes, then aligning the albumin-coding nucleotide sequences against the amino acid sequences for the haemoglobin genes will generate unpredictable results. In our case, both conditions are met. So just click **Yes** to proceed.

A standard File/Open dialog box appears. Double click the nucleotide sequence file that you have previously saved (e.g., **invert.fas**) and the nucleotide sequences will be aligned against the aligned amino acid sequences. These sequences are now ready for comparative sequence analysis. Save the aligned nucleotide sequences. These sequences will now be used to illustrate the point that sequence similarity becomes dependent on similarity in nucleotide frequencies when the sequences approach substitution saturation.

4. DATA ANALYSIS

Click **Sequences|Work on codon position 3**. DAMBE will now pick up all nucleotides at the third codon position and form a set of new sequences. You will not that the new sequences are just one third of the length of the original sequences. The reason for using only the third codon position is because this codon position is the most likely to experience substitution saturation among the three codon positions, and is therefore better suited for illustrating the effect of nucleotide frequencies on phylogenetic analysis.

Click **Seq. Analysis|Nucleotide Frequencies**. In the next dialog box, click the **Add All** button, and then click the **Go!** button. Nucleotide frequencies for each of the 21 sequences will be generated. Go directly to the very last part of the output where we get a summary of nucleotide frequencies shown in Table 1. Three species have been shown in bold in Table 1 because of their similarity in nucleotide frequencies. They have the lowest A and T frequencies, the highest frequency of C, and a relatively high frequency of G. With substitution saturation, these three sequences are expected to cluster together, to the exclusion of their respective sister taxa that do not have similar nucleotide frequencies.

Table 1. Testing heterogeneity of the nucleotide frequencies at the third codon positions of the 16 EF-1α sequences. $X^2 = 177.5$, DF = 45, P < 0.001. The four species with relatively higher GC and lower AT content than other s are shown in bold.

Species	A	C	G	T
Ann90063	0.1157	0.314	0.2397	0.3306
Ann90064	0.1077	0.3343	0.2403	0.3177
Bra90058	**0.0962**	**0.4835**	**0.2088**	**0.2115**
BraASEF1A	0.1951	0.25	0.1648	0.3901
Che90045	0.2912	0.217	0.1593	0.3324
Che90047	0.1401	0.3407	0.239	0.2802
Che90048	0.2775	0.2665	0.2198	0.2363
Che90051	0.174	0.2155	0.2072	0.4033
Che90052	0.2527	0.1566	0.1648	0.4258
Che90060	0.1456	0.2885	0.2527	0.3132
Hex90054	0.1566	0.2308	0.2225	0.3901
Hex90056	0.2424	0.1736	0.2094	0.3747
Hex90059	0.1978	0.3022	0.1621	0.3379
Mal90046	0.2687	0.2604	0.1025	0.3684
Mal90050	**0.0716**	**0.4683**	**0.314**	**0.146**
Mol90061	0.2247	0.2669	0.177	0.3315
Mol90062	0.2149	0.2479	0.2287	0.3085
Myr90049	0.1538	0.3489	0.217	0.2802
Myr90053	0.1551	0.3241	0.2438	0.277
Myr90055	**0.1083**	**0.4389**	**0.2917**	**0.1611**
Myr90057	0.1846	0.3003	0.2121	0.303

To perform a simplest phylogenetic analysis, click **Phylogenetics|Distance methods|Nucleotide** sequences. In the next dialog box, choose **K80** as the genetic distance (K80 refers to Kimura's two-parameter distance). Click the **Done** button. A neighbor-joining tree (Fig. 6), from genetic distances based on the K80 model, supported the expectation that species of similar nucleotide frequencies will be grouped together.

Two things are obvious. First, the tree is absurd, which is what we would have expected if all historical information has already been completely obliterated by substitution saturation. Second, the three sequences having similar nucleotide frequencies and shown in bold in Table 1 form a monophyletic taxon to the exclusion of their respective sister taxa.

All genetic distances that either ignore the frequency parameters (e.g., JC69 and K80), or ignore the transition bias (e.g., TN84) will group these three taxa together. If you try to use the maximum parsimony (MP) method, then two MP trees will be found, both grouping the three species to the exclusion of their true sister taxa. This is what we would have expected because the maximum parsimony method pays no attention to nucleotide frequencies or transition/transversion bias.

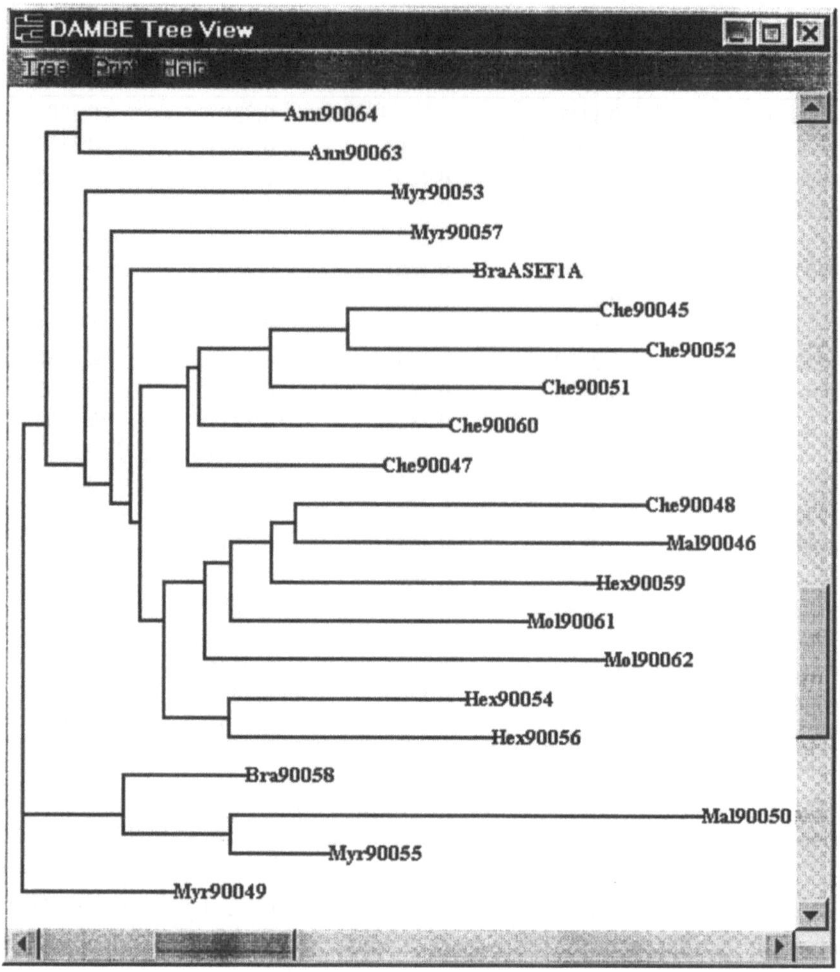

Figure 6. A neighbor-joining tree from the EF-1α sequences.

Compare the tree topology in fig. 6 with that in fig. 5. Should you include the third codon position in your phylogenetic analysis when sequences have experienced a long divergence time and consequently substitution saturation? The tree in fig. 5 is based on amino acid sequences. What would the tree be like if we use only the first and second codon positions in a phylogenetic reconstruction?

Chapter 8

Factors Affecting Codon Frequencies and Codon Usage Bias

1. INTRODUCTION

An amino acid is specified by a codon, and a protein is specified by a codon sequence. Because of the degeneracy of genetic codes, an amino acid is typically coded by several synonymous codons, which constitute what we call a synonymous codon family. For example, the synonymous codon family for the amino acid glycine is made of four codons: GGA, GGC, GGG, GGU.

Synonymous codons are not equivalent and there is often preferential use of one over the other. For example, the glycine is predominantly coded by GGC and GGU in the *E. coli* genome, but by GGA in mammalian mitochondrial genome. Why should there be such a difference?

It is now well known that genomes from distantly related organisms often exhibit different patterns of synonymous codon usage (Grantham et al. 1981; Grantham et al. 1980). In addition to this inter-genome difference, there are substantial inter-gene differences within the same genome (Gouy and Gautier 1982; Ikemura 1985; Ikemura 1992; Sharp et al. 1988; Sharp and Li 1987; Sharp and Mosurski 1986). How is this diversity of codon usage generated and maintained?

I will attempt to give you some possible answers to these questions, but first of all, we have to learn how to quantify the codon usage of genes or genomes so that comparisons can be made. Simply put, codon usage is characterized by the frequencies of the 64 codons. A comparison of codon usage among genes or genomes is simply a comparison of codon frequency tables.

There is another reason for studying codon frequencies. Recall that we need to know both the frequency parameters and rate ratio parameters in order to understand the substitution process. In previous chapters, we have briefly mentioned nucleotide-based models, which could have no more than three frequency parameters to estimate. You will learn later that nucleotide-based models are insufficient to describe evolution of protein-coding genes and that we need to have codon-based models which are much more complicated. There are 64 codons and consequently 63 frequency parameters to estimate. Ignoring termination codons still leaves us with about 60 frequency parameters. These frequency parameters are typically quite different from each other, and we need to appreciate the variation among them in order to understand the substitution process involving codons. It is these frequency parameters that we will examine in detail in this chapter. The rate ratio parameters of codon-based models will be left to later chapters.

If you find it outrageous that I should bother you with such a simple task of generating codon frequency tables, consider a Chinese proverb saying that "Familiarity begets insights". By forcing you to generate codon frequency tables and to stare at them, you will soon be able to discern subtle difference in codon usage patterns between genes and genomes.

2. GENERATING CODON USAGE TABLE WITH DAMBE

Start DAMBE, and open a file containing protein-coding nucleotide sequences. Click **Analysis|Codon Frequency**. A dialog box appears (Fig. 1). There are two lists in the dialog box. The one on the left lists the sequences that are available for selection. The one on the right lists sequences selected for computing codon frequencies. At this moment, the list on the right is empty.

- To select a single sequence, just click to highlight it, and then click the → button to move it to the right.
- To select neighboring sequences, click the first of the neighboring sequences to highlight it and then, holding down the shift key, click the last of the neighboring sequences. All the neighboring sequences will then be highlighted. Click the → button to move them to the right.
- To selection disjoint sequences, click each sequence while holding down the Ctrl key, and then click the → button to move the highlighted sequences to the right.

– If you have moved an unwanted sequence to the right, don't worry. Just highlight the sequence and click the ← button to move it back to the left.

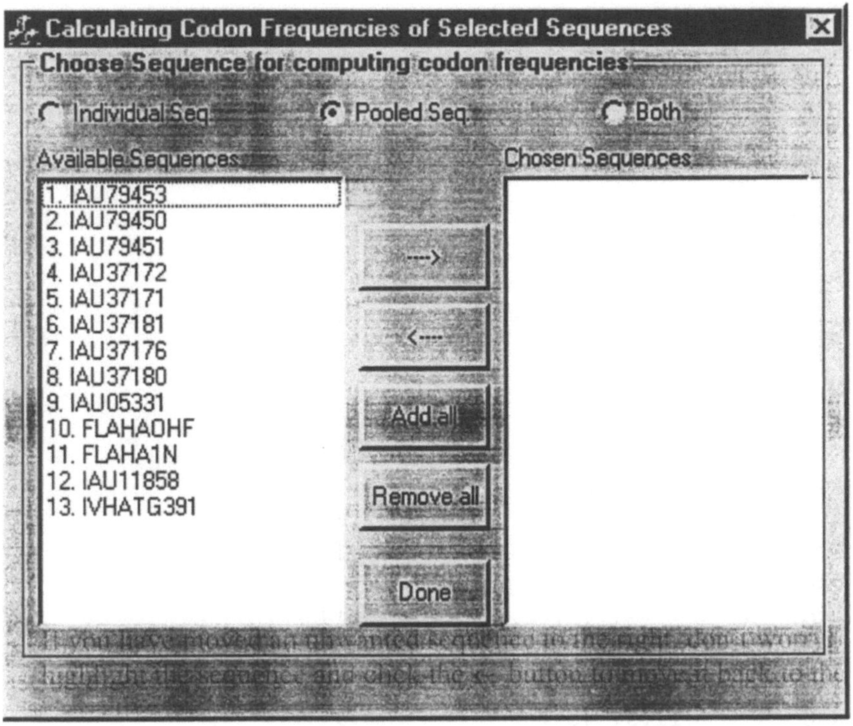

Figure 1. Dialog box for selecting sequences for generating codon frequency data. Note the radio buttons at the top of the dialog box

There are three options (radio buttons) on the top of the dialog box (Fig. 1). Clicking the option labelled **Individual sequence** will generate codon frequencies for each of the selected sequences. Clicking the option **Pooled** will generate codon frequencies for all selected sequences pooled together, i.e., you will get just one table of pooled codon frequencies no matter how many sequences you have selected. The third option, labelled **Both**, is for outputting codon frequencies of individual sequences as well as the pooled codon frequency table. The default is **Pooled**.

Once you have finished your selection, click the **Done** button. After a few seconds, the standard **file/save** dialog box appears. Type in the file name for saving the result, or simply use the default. Click the **Save** button. The file is saved in text format, and also automatically displayed on screen. A part of a default sample output, based on a segment of the Influenza A viruses, is shown below:

```
Output from sequences in file C:\MS\virus\virus.rst on
Tuesday, March 17, 1998

Sequence length =  969  (After excluding '?', '-' and 'n'.)
Number of codons =  323

From pooled sequences

AA   Codon     Mean Number(Sum)      RSCU
===================================================
A    GCA          8   ( 32 )          1.87
     GCC          3.3( 13 )            .75
     GCG          2.3( 9 )             .56
     GCU          3.5( 14 )            .82
R    AGA          6.3( 25 )           3.11
     AGG          4.8( 19 )           2.34
     CGA           .3( 1 )             .17
     CGC          0   ( 0 )           0
     CGG           .3( 1 )             .17
     CGU           .5( 2 )             .22
N    AAC         13.8( 55 )           1.1
     AAU         11.3( 45 )            .9
D    GAC          7.5( 30 )           1.45
     GAU          3.3( 13 )            .55
C    UGC          4.8( 19 )           1.08
     UGU          4   ( 16 )           .92
E    GAA         12.5( 50 )           1.34
     GAG          6.3( 25 )            .66
Q    CAA          6.3( 25 )           1.43
     CAG          2.5( 10 )            .57
G    GGA         12.3( 49 )           2.18
     GGC          2.3( 9 )             .41
     GGG          5.3( 21 )            .94
     GGU          2.8( 11 )            .47
H    CAC          3.8( 15 )            .82
     CAU          5.5( 22 )           1.18
I    AUA          9.5( 38 )           1.57
     AUC          4.8( 19 )            .81
     AUU          3.8( 15 )            .62
L    CUA          6.3( 25 )           1.46
     CUC          2.8( 11 )            .61
     CUG          6.5( 26 )           1.5
     CUU          3   ( 12 )           .7
```

	UUA	3.8(15)	.85
	UUG	3.8(15)	.88
K	AAA	12.3(49)	1.27
	AAG	7 (28)	.73
M	AUG	2.3(9)	1
F	UUC	5.5(22)	1.1
	UUU	4.5(18)	.9
P	CCA	9 (36)	2.02
	CCC	5 (20)	1.13
	CCG	1 (4)	.23
	CCU	2.8(11)	.63
S	AGC	4.5(18)	.85
	AGU	5.5(22)	1.02
	UCA	10.8(43)	2.02
	UCC	4.8(19)	.89
	UCG	1.3(5)	.23
	UCU	5.3(21)	.99
T	ACA	11.8(47)	2.09
	ACC	5.3(21)	.93
	ACG	2 (8)	.36
	ACU	3.8(15)	.63
W	UGG	6.8(27)	1
Y	UAC	8 (32)	1.05
	UAU	7.3(29)	.95
V	GUA	6.5(26)	1.29
	GUC	4 (16)	.8
	GUG	5.3(21)	1.04
	GUU	4.5(18)	.88

The codon usage table is based on the following sequences:

1 FLAHAOHF
2 FLAHA1N
3 IAU11858
4 IVHATG391

CodSite		A	C	G	U	Sum
1	Freq.	433	221	357	281	1292
	Prop.	.34	.17	.28	.22	1
2	Freq.	428	318	240	306	1292
	Prop.	.33	.25	.19	.24	1
3	Freq.	461	319	228	284	1292
	Prop.	.36	.25	.18	.22	1

The output is in two parts. The first is a table of codon frequencies categorized into codon families, and the second lists nucleotide frequencies separately for each of the three codon positions designated as **CodSite** in the output. There will be no second part if the sequences are not protein-coding nucleotide sequences. The 20 amino acids are designated by the one-letter notation (Table 1).

Table 1. One-letter and three-letter notation of the 20 amino acids.

Ala	A	Alanine	Met	M	Methionine
Cys	C	Cysteine	Asp	N	Asparagine
Asp	D	Aspartate	Pro	P	Proline
Glu	E	Glutamate	Glu	Q	Glutamine
Phe	F	Phenylalanine	Arg	R	Arginine
Gly	G	Glycine	Ser	S	Serine
His	H	Histidine	Thr	T	Threonine
Ile	I	Isoleucine	Val	V	Valine
Lys	K	Lysine	Try	W	Tryptophan
Leu	L	Leucine	Tyr	Y	Tyrosine

Below I will show you two interesting patterns as well as their possible interpretations. You are encouraged to discover other patterns that I have failed to discern.

3. DNA METHYLATION AND USAGE OF ARGININE CODONS

One obvious pattern from the codon frequency table above is the abundance of AGR codons relative to CGN codons, both coding for amino acid arginine. This can be interpreted as an evolutionary consequence of DNA methylation. DNA methylation is a ubiquitous biochemical process observed in both prokaryotes (Noyer-Weidner and Trautner 1993) and eukaryotes (Antequera and Bird 1993). In vertebrates, DNA methylation mainly involves the methylation of C in the CpG dinucleotide, which greatly elevates the mutation rate of C to T through spontaneous deamination of the resultant 5-methylcytosine (Barker et al. 1984; Cooper and Krawczak 1989; Cooper and Krawczak 1990; Cooper and Schmidtke 1984; Cooper and Youssoufian 1988; Ehrlich 1986; Ehrlich et al. 1990; Rideout et al. 1990; Schaaper et al. 1986; Sved and Bird 1990; Wiebauer et al. 1993). The elevated mutation pressure implies that a gene with many CpG dinucleotides would be unreliable. Thus, if an arginine is needed, then the protein gene is better off to have the arginine coded by AGR codons rather than by CGN codons. This is a selectionist explanation, i.e., it is evolutionarily more advantageous for the genome to code arginine by the AGR codon than by the synonymous CGN codons.

Because the gene the produced the result above is from a virus infecting mammalian species, i.e., the viral DNA will also be heavily methylated as its host DNA, we expect a high AGR usage relative to CGN usage for coding arginine.

For a quick comparison, I have list below a partial output for the EF-1α gene from the seven invertebrate species in the file **invert.fas** that comes with DAMBE. The level of DNA methylation is very low in invertebrate species. If the above interpretation is correct, then we should expect the relative frequency of AGR codons to CGN codons to be less AGR-biased in the invertebrate species than that observed in the gene from a virus infecting mammalian species. This is true. About 92% of arginine codons are AGR in the former, whereas only about 32% of arginine codons are AGR in the latter.

AA	Codon	Mean Number(Sum)	RSCU
R	AGA	2.9(20)	1.25
	AGG	1.6(11)	0.65
	CGA	0.7(5)	0.31
	CGC	0.6(4)	0.24
	CGG	0.4(3)	0.18
	CGU	7.9(55)	3.37

I hope that by this time you should come up with two criticisms for the arguments presented above. First, the comparison is between two totally different genes, which is the same as comparing apples and oranges. It is highly likely that the ratio of AGR codons over CGN codons is different among genes, even within the same genome. A valid comparison should be done either by many comparisons between homologous genes, or by taking a random (representative) sample of codons from heavily methylated and lightly methylated genomes. Second, even if such a valid comparison is made and the result is consistent with the pattern shown above, the selectionist interpretation is not valid because we can explain the pattern equally well by invoking mutation along. For example, if we start with 100 arginine codons, coded by 50 AGR codons and 50 CGN codons, i.e., the original ratio of AGR codons over CGN codons is 1. Now methylation will increase the mutation rate of CGN codons to TGN codons, which will obviously increase the ratio of AGR codons over CGN codons. Can you come up with a test that can distinguish between the mutationist explanation and the selectionist explanation?

4. TRANSCRIPTION EFFICIENCY AND CODON USAGE BIAS

Another pattern from the first part that is immediately obvious and interesting is that, for almost all codon families, the A-ending codon is used most often. This answers one question that we raised before. Let me refresh your memory just in case you forget. When we analyzed nucleotide frequencies of the same set of sequences, we noticed that A is the most frequent nucleotide in the Influenza A viruses. We proposed two possibilities for the preponderance of A. First, it might be caused by the preponderance of A-rich codons, e.g., Asn and Lys (which are coded by AAN codons), in the protein-coding gene. Alternatively, it might be caused by codon usage bias favoring A-ending codons for more efficient transcription (Xia 1996). By comparing the codon frequencies of the viruses with those from their hosts, we can conclude that both possibilities are likely true. The protein-coding genes in the viruses use significantly more amino acids coded by A-rich codons than those from the hosts, and the A-ending codon is almost invariably the more preferred codon in the viruses than in the hosts.

The over-use of A in the viral genes would increase the transcription efficiency because it is well known that ATP is much more abundant in the cellular medium than any other ribonucleotides (Bridger and Henderson 1983, pp. 4-5). It would be evolutionarily disadvantageous if the viral genes need a lot of rare ribonucleotides for transcription. This interpretation of codon usage bias has been termed the transcription hypothesis of codon usage, or THCU (Xia 1996) , which is featured in the first of the two case studies following this chapter.

5. TRANSLATIONAL EFFICIENCY AND CODON USAGE BIAS

We were briefly introduced to the translational efficiency hypothesis of codon usage bias in the previous section. To better appreciate the hypothesis, let us compile a codon frequency table from an *E. coli* gene, *gro*EL that codes for a protein affecting mRNA stability and is known to be highly expressed (Ikemura 1992).

Start DAMBE if you have not done so. Click **File|Read sequences from GenBank**. In the ensuing dialog box, enter **ECOGROELA** (which is the LOCUS name for the DNA sequence of the *gro*EL gene) into the textbox. Click the **Retrieve** button. In the next dialog box, click the **CDS** option button and then the **Proceed** button to get the coding sequences. In the next dialog box with five columns, click the LOCUS name, i.e., **ECOGROELA** in the first column. In the last column of the dialog box, we see that the coding sequence starts at position 28, with an ATG codon, which codes for methionine and is the initiation codon for initiating translation. The coding sequence ends at position 477, with a TGA codon, which is the termination codon signalling the end of the translation. In the bottom panel, the whole sequence is displayed with the coding sequence highlighted in red. Click the **Splice** button and then the **Done** button. The coding sequence will be extracted and displayed in DAMBE's display window. Save the sequence to a file in FASTA format.

To obtain a codon usage table, just click **Seq. Analysis|Codon frequency**. Click the **Individual** option button because we have just a single sequence. Click the sequence name in the left listbox and then the → button. Click the **Done** button and a table of codon frequencies is displayed.

Take a look at the glycine codons. Is the A-ending codon used most frequently among the four synonymous codons? Is the observed data consistent with the transcription efficiency hypothesis of codon usage? Can you offer an alternative hypothesis to explain the codon pattern in *E. coli*?

About 20 years ago, a Japanese biologist named Ikemura was asking himself the same set of questions, and he proposed what is now known as the translational efficiency hypothesis. The hypothesis states that there is strong selection favouring increased rate of protein synthesis and that a coding strategy that increases the rate of translation initiation and peptide elongation (and consequently increases the rate of protein synthesis) is favoured by natural selection. But what coding strategy would maximize translational efficiency?

The efficiency of translation depends mainly on the concentration of three chemical components participating in the translational process: the ribosome where translation takes place, the 20 amino acids that are the building blocks, and tRNA molecules that carry amino acids to the translation site. These tRNA molecules each have an anticodon to pair with a codon. For example, a glycine-carrying tRNA in *Escherichia coli* and *Salmonella typhimurium* may have an anticodon of CCA to pair with GGU, and another glycine-carrying tRNA may have an anticodon CCC to pair with GGG. In short, different tRNA molecules carrying the same amino acid may recognize different synonymous codons.

Suppose we have a GGG codon to be translated into glycine, but most glycine-carrying tRNA molecules in the cytoplasm recognize the synonymous GGU codon, but not the GGG codon. This implies that the GGG codon will, on average, be translated slowly. In contrast, a GGU codon is expected to be translated relatively fast. To maximize translational efficiency, and consequently the rate of biosynthesis, the synonymous codon usage should be biased towards the codon recognized by the most abundant tRNA species.

Ikemura did not stop at the hypothesis, but instead proceeded to test the hypothesis by measuring the concentration of various tRNA species and correlating it with codon usage. An elementary introduction to subsequent theoretical elaboration of codon usage and translational efficiency can be found in Xia (1998a). A more extensive and lucidly written review is also available (Akashi and Eyre Walker 1998). Case study 3 illustrates the basic ideas of the existing theory concerning codon usage bias.

6. CODON FREQUENCY AND THE MEAN AND VARIANCE OF PEPTIDE LENGTH IN ANCIENT PROTEINS

The frequency of the 64 trinucleotides in the primitive RNA or DNA sequences, before the genetic code is fixed, is expected to be $P_i \bullet P_j \bullet P_k$, with P_i, P_j, and P_k being the frequency of the nucleotide i, j and k (i, j, k = {A, C, G, T or U}) in the RNA or DNA sequences. This implies that, at the fixation of the genetic code, the ancient peptide-coding genes should have codon frequencies depending only on nucleotide frequencies. This has strong implications on the mean and variance of the length of ancient peptides.

If we take the universal genetic code for example and assume that the four nucleotide occurs in equal frequencies in the ancestral DNA or RNA sequences, then the stop codon is expected to occur once in about every 20 triplet codons, with the consequence that the peptide is, on average, just about 20 amino acids long. The distribution of the peptide length (l) follows the geometric distribution:

$$P(L = l \mid p) = p(1 - p)^l, \qquad (8.1)$$

where p is the proportion of stop codons and is equal to 3/64 for the universal genetic code with the assumption of equal codon usage. Note that an mRNA made of L codons (including the termination codon) will generate a peptide length of only L-1 amino acid residuals. The expected mean and variance of the ancient peptide length, assuming equal codon usage, are then

$$E(L) = \frac{1}{p},$$
$$Var(L) = \frac{(1-p)}{p^2} \qquad (8.2)$$

For the universal genetic code with p = 3/64, the expected mean and variance of the peptide length are 21.3 and 433.8, respectively, with less than 5% of the sequences longer than 120 amino acids.

By consulting the universal genetic code, we find that stop codons are coded by TAA, TAG and TGA. If we designate the four nucleotide frequencies by P_A, P_C, P_G, and P_T, then p in equation (8.2) equals ($P_tP_aP_a$ + $P_tP_aP_g$ + $P_tP_gP_a$). Noting that termination codons are AT-rich, we can immediately make one general prediction. If the frequency of T and A increases, then the frequency of stop codons will increase as a consequence, and the peptides should get shorter. This leads to three subpredictions. First, proteins in AT-rich genomes should be shorter than those in AT-poor genomes. Second, in vertebrate genomes with AT-rich and AT-poor isochores, we should expect proteins coded in AT-rich isochores to be shorter than those coded in AT-poor isochores. Third, mutations favouring T and A would lead to shorter peptides.

So far we have derived several predictions by completely ignoring the effect of selection. It is important to realize that, to understand the effect of selection, we need to know what would be the case in the lack of selection. Once we have recognized the pattern without selection, then any deviation from this pattern is likely to be caused by selection.

Let us review the predictions that we have so far derived. First, we have predicted, assuming no selection against stop codons, the mean peptide length is 1/p, and the associated variance is $(1\text{-}p)/p^2$, where p is the frequency of stop codons and can be expressed as ($P_tP_aP_a$ + $P_tP_aP_g$ + $P_tP_gP_a$). We further predicted that, in AT-rich DNA, p should be large and proteins, on average, should be shorter than those in AT-poor DNA. These predictions have never been tested.

In GenBank, there are many DNA sequences from a variety of genomes with very different AT content. There are a number of human protein-coding genes from GC-rich isochores and also many from AT-rich isochores. The former is expected to produce longer peptides than the latter. Can you compile a number of protein-coding genes to see if this prediction is favoured by empirical evidence?

Chapter 9

Case Study 2

Transcription Efficiency and Codon Usage Bias

1. INTRODUCTION

Genomes from distantly related organisms exhibit different patterns of synonymous codon usage (Grantham et al. 1981; Grantham et al. 1980). In addition to this inter-genome difference, there are substantial inter-gene differences within the same genome (Gouy and Gautier 1982; Ikemura 1985; Ikemura 1992; Sharp et al. 1988; Sharp and Li 1986; Sharp and Li 1987). Natural selection for increased translational efficiency has been proposed as the major hypothesis for the inter-genome and inter-gene differences in codon usage (Bulmer 1991; Kimura 1983; Kurland 1987a; Kurland 1987b; Robinson 1984; Xia 1998a). Three lines of evidence appear to support this hypothesis. First, the frequency of codon usage is positively correlated with tRNA availability (Gouy and Gautier 1982; Ikemura 1981; Ikemura 1982; Ikemura 1985; Ikemura 1992; Ikemura and Ozeki 1983). Second, the degree of codon usage bias is related to the level of gene expression, with highly expressed genes exhibiting greater codon bias than lowly expressed genes (Bennetzen and Hall. 1982; Sharp et al. 1988; Sharp and Devine 1989). Third, mRNA consisting of preferred codons is translated faster than mRNA artificially modified to contain rare codons (Sorensen et al. 1989).

Not only are there differences in codon usage bias among genomes and among genes within the same genome, but there are also differences in codon usage among different regions of the same gene. For example, gene regions of greater amino acid conservation tend to exhibit more dramatic codon usage bias than do regions of less amino acid conservation (Akashi 1994a). This has been proposed as resulting from selection for increased translational accuracy (Akashi 1994a; Hartl et al. 1994b), because selection for maximum translational efficiency does not seem satisfactory to explain the phenomenon. However, this can be accommodated by the translational efficiency hypothesis if one defines what is maximized as the rate of production of *correctly* translated proteins.

What all these studies have shown is that there is strong selection favouring increased rate of protein synthesis and that a coding strategy that increases the peptide elongation rate (and consequently increases the rate of protein synthesis) is favoured by natural selection. However, efficient protein synthesis depends not only on the peptide elongation rate, but also on the initiation rate. Moreover, four lines of evidence support the claim that the initiation of protein synthesis, rather than elongation of the peptide chain, is rate-limiting (Bulmer 1991). Thus, if there is selection for increased rate of protein synthesis, then we should expect selection to favour an increase of not only elongation rate, but also initiation rate. The evolutionary consequence of selection for increased elongation rate has been investigated and empirically documented extensively (Bulmer 1991; Gouy and Gautier 1982; Ikemura 1985; Ikemura 1992; Sharp et al. 1988; Sharp and Devine 1989; Sharp and Li 1986; Sharp and Li 1987; Xia 1998a). In contrast, the evolutionary consequence of selection for increased initiation rate has not been equally well studied.

The initiation rate is directly proportional to the encountering rate between mRNA molecules and ribosomes, and this encountering rate depends on the concentration of mRNA and ribosomes. Thus, the initiation rate of protein synthesis can be efficiently increased by increasing mRNA concentration. Both theoretical reasoning and empirical evidence suggest that the number of mRNA copies available is a rate-limiting factor in protein synthesis (Xia 1995). It is conceivable that natural selection should favour increased rate of transcription, and that a coding strategy leading to increased transcriptional efficiency should be at a selective advantage. Thus, studying the pattern of codon usage from the perspective of transcription adds one more dimension to our understanding of the evolution of genetic information.

2. CONSEQUENCES OF MAXIMIZING TRANSCRIPTIONAL EFFICIENCY

I here present an optimality model showing the effect of maximizing transcription rate on codon usage bias. Suppose that an mRNA molecule of length L is composed of A, C, G, and U with frequencies N_A, N_C, N_G and N_U, respectively ($N_A + N_C + N_G + N_U = L$). In terms of a chemical equation,

$$N_A A + N_C C + N_G G + N_U U \xrightarrow{k} mRNA \quad (9.1)$$

where k is the velocity constant of the transcriptional process. Let C be the concentration of transcribed mRNA, and let C_A, C_C, C_G and C_U be the concentration of A, C, G, and U, respectively, in the cellular matrix surrounding the active transcription site. Then, according to the law of mass action, the rate of transcription is

$$\frac{dC}{dt} = k\, C_A^{\,N_A}\ C_C^{\,N_C}\ C_G^{\,N_G}\ C_U^{\,N_U} \quad (9.2)$$

Evidently, if C_A is greater than C_C, C_G and C_U, then the transcription rate is increased by increasing N_A and decreasing N_C, N_G and N_U, with the constraint that $\sum N_i = L$, where i = A, C, G, U. Consequently, the maximum transcription rate is reached when $N_A = L$ and $N_C = N_G = N_U = 0$.

Equation (9.2) links the nucleotide composition of mRNA, i.e., N_A, N_C, N_G and N_U, to the relative nucleotide concentration in the cellular matrix at the transcription site, i.e., C_A, C_C, C_G and C_U. The equation predicts that the most frequently used nucleotide in mRNA molecules should be the same as the most abundant nucleotide in the cellular matrix. This implies that the relative concentration of the four nucleotides in the cellular matrix can affect patterns of synonymous codon usage. This hypothesis will hereafter be referred to as the transcription hypothesis of codon usage (THCU).

The same conclusion can be derived from a deterministic model with more explicit assumptions. Consider the time required to transcribe a single nucleotide i. Let r be the rate of nucleotides diffusing to the transcription site and P_i be the probability that the arriving nucleotide is nucleotide i. Note that P_i (where i = A, C, G, U) simply represents the relative availability of the four nucleotides. Let t_l be the time spent in linking this nucleotide to the elongating mRNA chain, and t_r be the time spent in rejecting each of the wrong nucleotides that diffuse to the transcription site prior to the arrival of the nucleotide i. Now the total time spent in transcribing nucleotide i is:

$$T_i = \frac{1}{r P_i} + t_l + \left(\frac{1}{P_i} - 1\right) t_r \quad (9.3)$$

where the first term on the right-hand side of the equation is the time needed for the correct nucleotide to arrive at the transcription site, and the third term represents time spent in rejecting the wrong nucleotides prior to the arrival of the correct nucleotide. The total time (T) required to transcribe L nucleotides (total elongation time) can be shown to be:

$$T = \sum_{i=1}^{4} N_i \qquad T_i = L(t_l - t_r) + \frac{1+rt_r}{r} Y \tag{9.4}$$

where

$$Y = \sum_{i=1}^{4} \frac{N_i}{P_i} = \frac{N_A}{P_A} + \frac{N_C}{P_C} + \frac{N_G}{P_G} + \frac{N_U}{P_U}$$

Note that N_i is a property of the mRNA, whereas P_i is a property of the cellular matrix.

Our objective, then, is to find the conditions that minimize T. Because t_l, t_r and L are not dependent on N_i and P_i, they are treated as constants. Thus, minimizing T in equation (9.4) is equivalent to minimizing Y. We rewrite Y as:

$$Y = \frac{N_A P_C P_G P_U + N_C P_A P_G P_U + N_G P_A P_C P_U + N_U P_A P_C P_G}{P_A P_C P_G P_U} \tag{9.5}$$

If P_A is the largest of the four, then ($P_C\, P_G\, P_U$) is smaller than either ($P_A\, P_G\, P_U$), ($P_A\, P_C\, P_U$) or ($P_A\, P_C\, P_G$). It is therefore obvious that minimization of Y in equation (9.5), given that P_A is the largest of the four, requires an increase in N_A and a decrease in N_C, N_G and N_U, with the minimum of Y reached when $N_A = L$ and $N_C = N_G = N_U = 0$. The general prediction from the optimality model, therefore, states that whenever different nucleotides in the cellular matrix differ in relative availability, the codon usage of the mRNA should evolve towards increasing the frequency of the most abundant nucleotide in the cellular matrix. Thus, we reached the same conclusion as that from the law of mass action.

Most synonymous codons differ at the third codon site. According to the general prediction above, we expect that, within each codon family, a codon ending with a nucleotide that is the most abundant in the cellular medium should be used the most frequently. This leads to three testable predictions (Predictions One to Three below). In addition, because introns are also transcribed and should be subject to selection maximizing transcription efficiency, we expect a nucleotide species to be used more frequently in introns when the concentration of that nucleotide species increases in the cellular medium (Prediction Four below).

3. PREDICTIONS AND EMPIRICAL TESTS

Prediction One: A-ending codons should be more frequent than alternative synonymous codons in mitochondrial protein genes: The concentration of cellular ATP is much higher than that of the other three nucleotides (C, G, and U), and the ATP concentration in mitochondria is still higher than that in cytosol (Bridger and Henderson 1983, pp. 4-5). The high ATP concentration in mitochondria might be caused by many factors, and one of these factors is that mitochondria have an efficient transport system to bring ADP into mitochondria for ATP production, but the transport system does not carry non-adenine nucleotides (Bridger and Henderson 1983; Olson 1986). Given that ATP concentration is higher than that of the other three nucleotides in mitochondria, we should expect synonymous codon usage to be biased toward A-ending codons in mitochondria to facilitate transcription, according to THCU.

How to test this prediction by directly accessing molecular data already available in GenBank? What we need is to get a few complete mitochondrial genomes, splice out the protein-coding sequences and obtain a codon table. You can hardly think of anything simpler.

Start DAMBE if you have not yet done so. Click **File|Read sequences from GenBank**. A dialog box (fig. 1) appears for you to choose relevant options that we have already discussed in detail in the chapter dealing with GenBank accessing. If you click the **General Search** option button and type in a search string such as "complete+mitochondrial+genome", you may get too many hits because there are now many mitochondrial genomes sequenced and deposited in GenBank. For example, there are now at least 25 completely sequenced mitochondrial genomes from mammalian species alone in GenBank.

For illustrating the test of the first prediction, we will just get a single complete mitochondrial genome from the cow, with accession number J01394. Click the top option button, labelled **Search GenBank with accession number or LOCUS name**, and type in J01394 in the text box. Leave everything as default as shown in fig. 1. Click the **Retrieve** button and the complete mitochondrial genome from the cow will be retrieved, which may take a minute or two depending on the speed of your network.

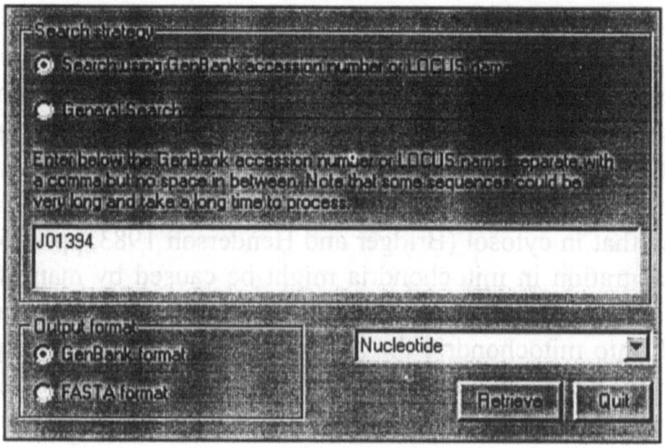

Figure 1. Obtain a complete mitochondrial genome for the cow, *Bos taurus*, from GenBank by using the accession number.

As we have learned in the chapter dealing with processing GenBank files, DAMBE can take advantage of the rich information contained in the FEATURES table of a GenBank file and splice specific gene segments, such as CDS, introns, rRNA, etc. This is why we specified the **GenBank** format rather than the **FASTA** format (lower left in fig. 1).

After you have retrieved the GenBank file, the following dialog box appears (fig. 2). The default is **Whole sequence**, which is not what we need. What we want to have are the coding sequences, i.e., CDS for the 13 protein-coding genes in the mitochondrial genome. So click the **CDS** option button and click the **Proceed** button. Another dialog box will appear, with the forth column (the second one from the right) showing the name of each protein-coding gene in the cow mitochondrial genome. There should be a total of 13 protein-coding genes in the mitochondrial genome. This forth column is for you to specify which gene you wish to extract from the lengthy genome. Just highlight all the genes in the forth column and click the **Splice** button. DAMBE will search through the FEATURES table in the GenBank file and extract all protein-coding genes from the GenBank file. Click the **Done** button to end the extraction.

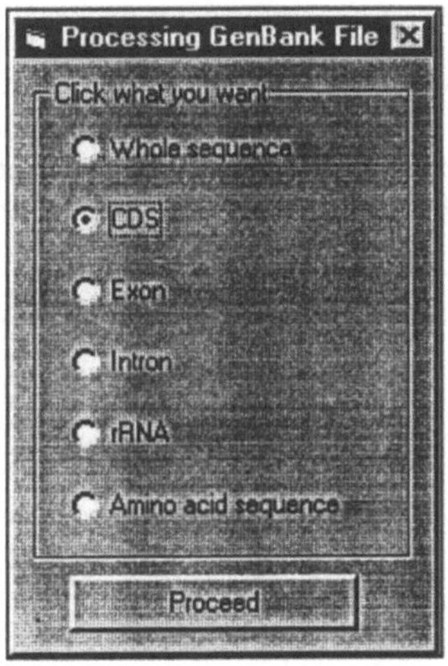

Figure 2. Dialog box for specifying which gene segments are needed for further study

You will be told that the genes are of different lengths, and asked if you wish to align the sequences. Click **No** because there is no point in aligning non-orthologous genes. All DNA sequences from the 13 protein-coding genes will be displayed in DAMBE's display window. To verify that the extraction is correct, click **Sequences|Working on amino acid sequences** to translate the protein-coding nucleotide sequences into amino acid sequences. The reason for doing this is not because DAMBE may make mistakes in extracting, but because some sequences in GenBank have been entered wrongly. For example, if the cytochrome-b sequence starts from position N and ends at position M of the complete mitochondrial sequence, the FEATURES table might specify the numbers wrong. For protein-coding genes, the worst thing that could happen is when N is wrongly by one or two nucleotides, which introduces an artefact of a frame-shifting mutation. Such mistakes will lead to several stop codons embedded in the nucleotide sequence, which will be translated into "*". If you see more than one "*" in the translated sequence, then you should check whether the FEATURES table has been entered properly.

If everything goes well, and all 13 protein-coding sequences are correctly extracted, then click **Seq. Analysis|Codon frequencies**. This produces codon frequencies and relative synonymous codon usage (RSCU), reformatted in Table 1. We note immediately the excess of A-ending codons (shown in bold type) in all synonymous codon families. The probability that RSCU for A-ending codons is not greater than 1 is less than 0.0001. The pattern is similar for mitochondrial genome from other mammalian species rat, rabbit, sheep, human, and macaque. The empirical data thus strongly support the first prediction.

Table 1. Codon usage bias in the mitochondrial genome of the cow, *Bos taurus*. Values are based on all protein-coding genes with a total of 3800 codons in the complete mitochondrial sequence (GenBank accession number J01394). AA = one letter code for amino acid; N = number of codons; RSCU = relative synonymous codon usage (calculated as the observed frequency of a codon divided by its expected frequency under the assumption of equal codon usage, Sharp et al. 1986); * = stop codons.

Codon	AA	N	RSCU	Codon	AA	N	RSCU
GCU	A	52	0.84	**CAA**	**Q**	**79**	**1.82**
GCC	A	91	1.47	CAG	Q	8	0.18
GCA	**A**	**103**	**1.66**	CGU	R	7	0.44
GCG	A	2	0.03	CGC	R	11	0.70
GAA	**E**	**78**	**1.64**	**CGA**	**R**	**42**	**2.67**
GAG	E	17	0.36	CGG	R	3	0.19
GGU	G	29	0.53	UCU	S	51	1.11
GGC	G	62	1.13	UCC	S	65	1.42
GGA	**G**	**97**	**1.77**	**UCA**	**S**	**99**	**2.16**
GGG	G	31	0.57	UCG	S	5	0.11
AAA	**K**	**90**	**1.78**	**UAA**	*****	**8**	**3.20**
AAG	K	11	0.22	UAG	*	1	0.40
UUA	**L**	**110**	**1.11**	**AGA**	*****	**1**	**0.40**
UUG	L	16	0.16	AGG	*	0	0.00
CUU	L	62	0.62	ACU	T	44	0.57
CUC	L	95	0.95	ACC	T	96	1.25
CUA	**L**	**285**	**2.86**	**ACA**	**T**	**153**	**1.99**
CUG	L	29	0.29	ACG	T	15	0.19
AUA	**M**	**218**	**1.66**	GUU	V	40	0.84
AUG	M	44	0.34	GUC	V	48	1.01
CCU	P	42	0.87	**GUA**	**V**	**87**	**1.83**
CCC	P	63	1.31	GUG	V	15	0.32
CCA	**P**	**85**	**1.76**	**UGA**	**W**	**92**	**1.77**
CCG	P	3	0.06	UGG	W	12	0.23

Prediction Two: The proportion of A-ending codons in each synonymous codon family should be smaller in nuclear protein genes than in mitochondrial protein genes: Whereas ATP concentration should be much higher than the concentration of non-ATP nucleotides in mitochondria for reasons stated in the previous paragraph, the difference in concentration between ATP and non-ATP nucleotides should be relatively small in the nucleus because ATP concentration is lower in nucleus than in mitochondria. ATP concentrations in rat liver cytosol and mitochondria were 6.2 ± 0.63 and 7.5 ± 0.73 μmol/(ml water), respectively (Bridger and Henderson 1983, p. 5). The actual difference is expected to be greater because mitochondrial preparation was not absolutely free of cytosol contamination and vice versa. It is believed that little difference exists in ATP concentration between nucleus and cytoplasm (Bridger and Henderson 1983, p. 5), i.e., the difference in ATP concentration between mitochondria and cytosol is also the difference between mitochondria and nucleus.

Given the lower concentration of ATP in nucleus than in mitochondria, we should expect A-ending codons to be less frequent in the nuclear genome than in the mitochondrial genome, which is also true (Table 2). Thus, the difference in synonymous codon usage between the nuclear genome and the mitochondrial genome appears to be explained, at least partially, by the difference in relative ATP concentration between the nuclear medium and the mitochondrial medium. Can you collect your own data to test this second prediction?

Table 2. Proportion of A-ending codons (N_A / N) in each synonymous codon family in the nuclear (NUC) and mitochondrial (mtDNA) genomes in the cow, *Bos taurus*. Data from *E. coli* were used for comparison. N_A = number of A-ending codons, N = total number of codons. Data from mitochondrial DNA were based on data in Table 1. Data for Arginine is limited to CGN codons (i.e., excluding AGA and AGG that are stop codons in mtDNA). Note the relative deficiency of A-ending codons in the nuclear genome relative to the mitochondrial genome in the cow ($P < 0.0001$ based on paired T-Test). The pattern is similar for rat, rabbit, cow, sheep, human, and macaque.

AA	Nuc	MtDNA	*E. coli*
Ala	0.1733	0.4153	0.22
Arg	0.1593	0.6667	0.06
Gln	0.2136	0.9080	0.31
Glu	0.3383	0.8211	0.70
Gly	0.2130	0.4429	0.09
Leu	0.0899	0.6616	0.14
Lys	0.3469	0.8911	0.76
Pro	0.2235	0.4404	0.20
Ser	0.1906	0.4500	0.12
Thr	0.2211	0.4968	0.12
Val	0.0766	0.4579	0.17

An alternative explanation for the difference between mtDNA and nuclear DNA in Table 2 is that the prokaryotic ancestor of mtDNA had a high frequency of A-ending codons, and that this high frequency of A-ending codons has been maintained through evolutionary inertia rather than through any optimization process suggested by THCU. If this is true, then we would expect prokaryotic genomes, which presumably share the same ancestor with the mitochondrial genome, also to exhibit a high frequency of A-ending codons. This expectation is clearly not fulfilled (Table 2). The frequency of A-ending codons in *E. coli* genome is significantly smaller ($p < 0.0001$, Table 2) than that of the mitochondrial genome in the cow. The pattern in Table 2 holds true if the cow in Table 2 is replaced by other eukaryotic organisms such as rat, rabbit, sheep, human, Macaca, Saccharomyces, or Drosophila. Note that the mtDNA and nuclear genome have diverged a very long time. So an explanation of evolutionary inertia is perhaps unnecessary in the first place.

Prediction Three: The proportion of A-ending codons should be greater in organisms with a high weight-specific metabolic rate (SMR) than in organisms with a low SMR: Different animal species differ greatly in SMR (measured as O_2 consumption in unit of $ml \cdot hr^{-1} \cdot g^{-1}$). In mammals, SMR is inversely correlated with body size, with the mouse having much higher SMR than the cow [Altman, 1972 #70, pp. 1613-1616). Differences in SMR among animals of different body sizes are correlated with the number of mitochondria per unit volume of tissue, with mammals of high SMR having more mitochondria per unit volume of tissue than mammals of small SMR (Ekert and Randall 1983, pp. 698-699; Mathieu 1981; Smith 1956). According to Weibel's (1984) authoritative review, the cell's potential for ATP production is proportional to the volume density of its mitochondria. This explains the rapid decrease of maximum sustainable metabolic rate with increasing body weight (decreasing volume density of mitochondria) in mammalian species (Hochachka 1991). In light of all these related lines of evidence, I think it reasonable to assume that nucleotide production is more ATP-biased in small mammalian species with a high SMR than in large mammalian species with a low SMR. In other words, the availability of cellular ATP (relative to the other three nucleotides, C, G, and U) is greater in small mammalian species with a high SMR than in large mammalian species with a low SMR.

If the inference above is correct, then we should expect a greater proportion of A-ending codons in small mammals, such as the mouse with SMR equal to 1.59, than in large mammals, such as the cow and sheep with SMR equal to 0.127 and 0.206, respectively (Altman and Dittmer 1972, pp. 1613-1616). We will now test this expectation empirically by using the erythropoietin gene from the mouse, cow and sheep. You may choose alternative genes to check for the generality of the conclusion derived from the erythropoietin gene alone.

The erythropoietin gene has already been sequenced for the cow, sheep and mouse, and the DNA sequences have been deposited in GenBank with accession numbers L41354, Z24681, and M12482. We will now read these sequences from GenBank.

Start DAMBE and click **File|Read sequences from GenBank**. When a dialog box (fig. 3) appears, just type in the three accession numbers in the text box and click the **Retrieve** button. The data retrieval may take from a few seconds to a few minutes, depending on the speed of your computer network. Bearing in mind that a sequence file retrieved from GenBank may contain not only protein-coding genes, but also RNA sequences, even for protein-coding genes, there are introns and untranslated 5' beginning or 3' ending sequences. The DNA sequence that specify the linear sequence of the protein is designated as CDS in GenBank files. Thus, when you are prompted for which kind of sequences you wish to extract from GenBank files, just click the **CDS** option button and then click the **Proceed** button.

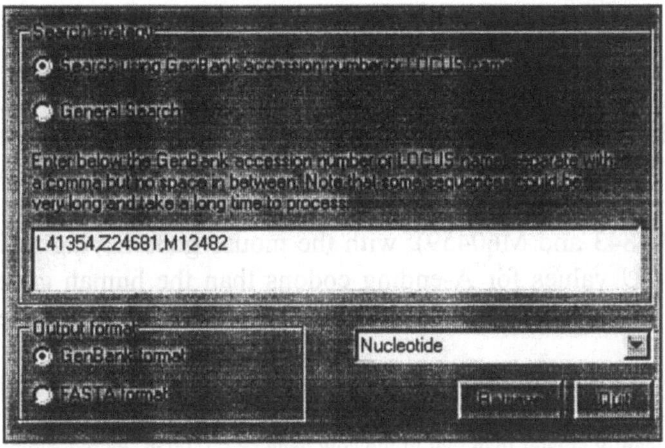

Figure 3. Retrieving sequences from GenBank by using GenBank accession numbers.

You will be presented with another dialog box, which has been explained in detail in the short chapter dealing with processing GenBank files, which you should consult if you intend to use GenBank files in the future. If you do not wish to consult that chapter, then just follow the instruction on the screen and extract the CDS sequences for the erythropoietin gene from the three mammalian species.

Once the CDS sequences have been extracted, you will be asked whether to align the sequences if the three sequences are not of equal length. If you click **Yes**, then the nucleotide sequences will be aligned. However, such alignment may introduce gaps of one or two nucleotides that are clearly artefacts. For protein-coding genes, it is always better to first translate the nucleotide sequences into amino acid sequences, align the amino acid sequences and then align the unaligned nucleotide sequences against aligned amino acid sequences. In short, when you are asked whether to align the sequences, click **No** instead, and the sequences will be displayed on screen.

Now first save the sequences to a file, say **erythro.fas**, in the FASTA format by clicking **File|Save as (converting)**. We will now translate the nucleotide sequences into amino acid sequences, Click **Sequence|Work on amino acid sequences**. DAMBE will automatically translate these nucleotide sequences into amino acid sequences. Click **Sequences|Align sequences with ClustalW** to align the sequences.

Once the amino acid sequences have been aligned, click **Sequences|Align Nuc. Seq. against aligned AA seq in the buffer**. When prompted for the input file name, just click the **erythro.fas** file that you have saved a moment ago. The nucleotide sequences will be aligned, and ready for further analysis.

Click **Seq. Analysis|Codon frequencies** to obtain the relative synonymous codon usage (RSCU) for the three genes. A large RSCU value means more frequent usage. Part of the output is summarized in Table 3. A-ending codons are used significantly more frequently in the mouse gene than in the cow and sheep gene. Complete DNA sequences are also available for the erythropoietin receptor gene from the mouse and human (accession numbers J04843 and M60459), with the mouse gene having a significantly greater RSCU values for A-ending codons than the human gene. SMR for the human is 0.228, which is much smaller than that for the mouse.

Table 3 Relative synonymous codon usage (RSCU) in the cow (Bos), sheep (Ovis) and mouse (Mus) erythropoietin gene (complete sequence). RSCU for A-ending codons in the mouse gene is significantly greater than that in the cow gene (P = 0.0056, Paired-sample T-Test) and that in the sheep gene (P = 0.0041). There is no difference between the cow gene and the sheep gene (P = 0.5451, Paired-sample T-Test).

Codon	AA	Bos	Ovis	Mus
UUA	L	0.00	0.00	0.34
CUA	L	0.19	0.18	0.51
AUA	I	0.00	0.43	0.75
GUA	V	0.00	0.00	0.33
UCA	S	0.50	0.50	1.09
CCA	P	2.22	2.18	2.13
ACA	T	0.44	0.44	1.00
GCA	A	0.80	0.70	0.57
CAA	Q	0.00	0.00	0.29
AAA	K	0.67	0.67	1.00
GAA	E	0.83	0.83	1.20
CGA	R	1.13	1.13	0.86
AGA	R	0.38	0.38	1.29
GGA	G	0.67	0.67	0.80

The test in Table 3 is weak for two reasons. First, the test is based on one gene. One cannot make generalizations based on one or few genes. To address this problem, I have compared codon usage between the mouse (SMR = 1.59) and the rat (SMR = 0.84) based on 877 and 833 genes, respectively, from the mouse and the rat. The mouse genes use A-ending codons significantly more frequently than the rat genes (P = 0.0006). These data strongly supported the third prediction.

The second problem with the test in Table 3 is that the mouse differs from the cow and sheep not only in metabolic rate, but also in many other ways, each of which could potentially be responsible for the difference in the usage of A-ending codons. One thing we can do to overcome this problem is to make more comparisons. I have also compared codon usage between the rabbit (SMR = 0.47) and the cow (SMR = 0.13) based on 133 and 261 genes, respectively, from the rabbit and the cow. There are 36417 and 58199 codons for the rabbit and the cow, respectively, of which 7614 and 11762 codons, respectively, are A-ending codons. A X^2-test showed that rabbit genes contain significantly more A-ending codons than the cow genes (X^2 = 6.699, DF = 1, P = 0.0096), which is again consistent with the third prediction. A similar comparison between the human (1952 genes) and the macaque (19 genes) did not show any significant difference, which is perhaps attributable to the small number of macaque genes and to the fact that the difference in SMR between the two species are not as great as that between the rat and the mouse or between the rabbit and the cow. In short, when two species differ much in SMR, they also differ in the use of A-ending codons in the direction predicted by THCU; when two species differ little in SMR, they also have similar codon usage.

You will be better equipped to test the predictions above after you have learned phylogenetic reconstruction and comparative methods.

Prediction Four: The A-content of introns should be greater in organisms with a high weight-specific metabolic rate (SMR) than in organisms with a low SMR: This prediction is interesting in two aspects. First, its confirmation would strengthen THCU. Secondly, it helps to distinguish between the transcription hypothesis and translational hypothesis concerning codon usage. The transcription hypothesis predicts that both introns and coding sequences should show the predicted "nucleotide usage bias", whereas the translation hypothesis predicts that only coding sequences should exhibit nucleotide usage bias (or codon usage bias).

The test of Prediction Four can be illustrated with the cytoplasmic β-actin gene, which has been sequenced for the human and the rat, with GenBank LOCUS names HUMACCYBB and RATACCYB, respectively. The gene from both the human and the rat contain five introns, which are spliced out, joined, and the percentage of A nucleotide calculated. Because the rat has a much high metabolic rate (0.84) than the human (0.23), Prediction Four would be supported if the introns of the rat gene have a higher percentage of A nucleotide than those of the human gene. Such a test should be applied to many genes to increase the generality of the test results.

I retrieved DNA sequences from GenBank by using proteins listed in Table 1 of chapter 4 in Li and Graur (1991) as keywords, which resulted in a total of 756 DNA sequences for various mammalian species. Many protein-coding sequences in GenBank do not contain sequence information on introns. Some genes have intron sequences for only one species, which are useless for our comparative purpose (which requires intron information from at least two species differing in metabolic rate, SMR). Some genes contain only partial intron sequences, which are discarded. Also discarded are those intron sequences with long stretches of unresolved sites, i.e., marked by "nnnnn.....". For the few genes that do contain complete intron sequences from multiple species, only five (skeletal α-actin, cytoplasmic β-actin, growth hormone, α- and β-globin genes) can have their exons and introns aligned properly for valid comparisons.

The percentage of A nucleotide in introns for each of the five genes representing multiple mammalian species was displayed in Table 4, together with the corresponding SMR values. Although the data are limited, we do find a consistent pattern for each of the five genes that the A-content of introns is greater in organisms with a high weight-specific metabolic rate (SMR) than in organisms with a low SMR (Table 4). For example, rodents have higher P_A than the human and the ungulates.

Table 4 Data from five protein genes for testing Prediction Four. LOCUS designates LOCUS name in GenBank. P_A - the proportion of nucleotides A. N_{total} - the total number of codons. Phylogenetically similar species are grouped next to each other. Note that species with higher SMR values tend to have higher A-content (P_A). All SMR (weight-specific metabolic rate) values are taken from Altman and Dittmer (1972, pp. 1613-1616). * - The reported value is a range (0.20 - 0.25).

Gene	Species	LOCUS	SMR	P_A	N_{total}
Skeletal muscle	Pig	SSU16368	0.126	0.1609	1243
α-actin	Cow	BTU02285	0.127	0.1579	1267
	Human	HUMSAACT	0.228	0.1539	1345
	Rat	RATACSKA	0.84	0.1954	1566
	Mouse	MUSACASA	1.59	0.2029	1498
Cytoplasmic	Human	HUMACCYBB	0.228	0.1099	1547
β-actin	Rat	RATACCYB	0.84	0.1533	1690
Growth	Pig	PIGGH	0.126	0.1726	927
Hormone	Cow	BOVGHGH	0.127	0.1715	974
	Sheep	SHPGHOV	0.200	0.1801	977
	Goat	GOTGHRA	0.233	0.1859	979
	Human	HUMGHN	0.228	0.2204	812
	Macaque	MMU02293	0.43	0.2212	764
	Rat	RATGROW2	0.84	0.2575	1275
α-Globin	Macaca	MACHBA	0.43	0.1226	261
	Mouse	MUSHBA	1.59	0.1992	256
β-Globin	Human	HUMBETGLOA	0.228	0.2776	980
	Mouse	MUSHBBH0	1.59	0.3016	1177
	Echidna	TGLHBB	0.225*	0.2242	611

4. AN ALTERNATIVE EXPLANATION

One alternative hypothesis for the patterns shown in Tables 1-4 is that of mutation bias. For example, a greater mutational pressure favouring A against G in the mitochondrial genomes than in the nuclear genomes would result in a greater proportion of A-ending codons in the mitochondrial genes than in the nuclear genes. Martin (1995) has argued that organisms of high metabolic rate should experience higher mutation rate favouring A than organisms of low metabolic rate.

The mutation hypothesis can be distinguished from THCU because the two hypotheses have different predictions. Let us first focus on the consequence of mutation favouring A against G. Suppose a protein gene with equal number of A, C, G, and T distributed randomly on both template and non-template strands (i.e., the original sequence in fig. 4). When five G's are replaced by five A's through mutation on the template strand, five C's will consequently be replaced by five T's on the non-template strand. Because mutation occurs randomly on both template and non-template strand of the gene, we also expect five G's to be replaced by five A's on the non-template strand and five C's to be consequently replaced by five T's on the template strand. The net result is that on either template or non-template strand, the increment in the number of A nucleotides (five in our fictitious case) is matched by the increment in the number of T nucleotides (also five in our fictitious case). In other words, $\Delta N_A = \Delta N_T$ on both template and non-template strands (fig. 4), so that A-ending codons and T-ending codons will be used equally frequently, and both used more frequently than G-ending and C-ending codons.

In contrast to the mutation hypothesis, THCU predicts that, with ATP more readily available than other nucleotides, the protein gene should evolve towards maximizing the use of A in mRNA (i.e., maximizing the number of A on the non-template strand of the coding sequence, fig. 4). This will result in an increase in the number of A, and a decrease in the number of T in the non-template strand of the gene (fig. 4). In short, although both THCU and the mutation hypothesis would predict that A-ending codons should be much more frequent than G-ending codons, the two hypotheses differ in that THCU predicts A-rich and T-poor on the non-template strand, whereas the mutation hypothesis (e.g., with mutation favouring A against G) predicts that both strands should be AT-rich and GC-poor, with A's and T's distributed equally on the two strand (fig. 4).

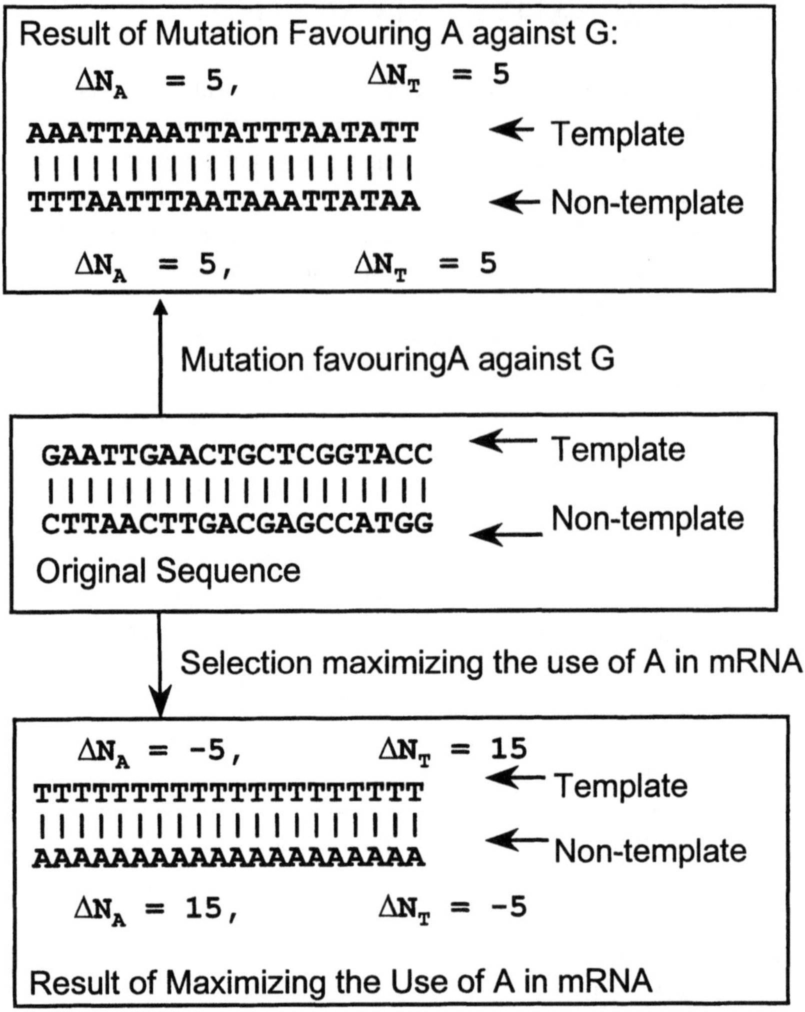

Figure 4. Contrasting predictions from the mutation hypothesis (with mutation favouring A against G) and THCU (transcription hypothesis of codon usage). THCU predicts A-richness and T-poorness in the non-template strand (bottom panel), and the opposite in the template strand. Δ designates increment. Because mutation favouring A against G is expected to occurred equally frequently on both DNA strands, the mutation hypothesis expects both DNA strands to accumulate equal number of A's and, consequently, equal number of T's, so that both strands will be AT-rich and GC-poor.

The mutation hypothesis seems to explain satisfactorily the pattern of codon usage in Drosophila mitochondrial DNA. The number of codons ending with A, C, G, and T in *Drosophila yakuba* is 1052, 107, 45, and 1092, respectively, for protein genes on the H-strand, and is 403, 6, 31, and 428, respectively, for protein genes on the L-strand. Thus, Drosophila mtDNA is AT-rich, with A-ending and T-ending codons used roughly equally, and both used much more frequently than C-ending and G-ending codons. These fulfil the prediction based on mutation hypothesis (top panel of fig. 4). In neither strand do we observe A-richness and T-poorness expected from THCU (fig. 4).

Additional evidence confirming that codon usage in Drosophila is mainly controlled by mutations favouring A or T comes from an AT-rich region flanking the origin of replication. This region spans 1.0-5.1 kb and is homologous in various Drosophila species (Fauron and Wolstenholme 1980a; Fauron and Wolstenholme 1980b; Goddard and Wolstenholme 1980). The region exhibits extensive sequence divergence, suggesting that the nucleotide sequence is mainly under the control of mutation bias (Goddard et al. 1982). The fact that the region is made of almost entirely of AT pairs implies that the mutation spectrum in Drosophila is strongly AT-biased, and that the preponderance of A-ending and T-ending codons in Drosophila mtDNA can be explained as a consequence of the mutation bias.

Another DNA region that appears to be strongly affected by mutation bias is the D-loop of mammalian mtDNA. Goddard *et al.* (1982) has suggested that the D-loop is homologous to the highly variable AT-rich region in Drosophila mtDNA mentioned above. Like the AT-rich region in Drosophila, the D-loop also flanks the replication origin, is also highly variable in nucleotide sequences (Avise et al. 1994), and is not transcribed except for perhaps a few bases. Thus, the nucleotide composition of the D-loop should reflect the mutation spectrum in the mammalian mtDNA. The number of A, C, G, and T in the mouse D-loop is 258, 104, 218, and 299, respectively. This is consistent with what we would expect if the D-loop is under mutation bias favouring A or T (top panel of fig. 4).

The mutation hypothesis, however, fails in explaining the pattern of codon usage in mammalian mtDNA. The data in Table 1 shows that A-ending codons are always much more frequently used than T (or U)-ending codons in bovine mtDNA, in contrast to what we see in Drosophila mtDNA where A-ending and T-ending codons are used equally frequently, and also in contrast to the D-loop region where T is more frequent than A.

The data from mouse mtDNA further highlight the inadequacy of the mutation hypothesis. The number of codons ending with A, C, G, and T in the mouse mtDNA is 1677, 1000, 117, and 825, respectively, with A-ending codons far outnumbering not only G-ending codons, but also T-ending codons. This pattern is the same as what we see in Table 1 for the cow and is expected from THCU, but not from the mutation hypothesis. (Note that there are more NNY codons than NNR codons in mammalian mtDNA, with the difference > 300. So the observed excess of A-ending codons and deficiency of T-ending codons in mammalian mtDNA is not a consequence of protein genes made of mostly NNR codons).

Although the pattern of codon usage in mtDNA is more satisfactorily explained by THCU than by the mutation hypothesis, one can still argue that the difference in codon usage between mtDNA and nuclear DNA (Table 2) is attributable to mutations in mtDNA more biased in favour of A than mutations in nuclear genome, which could result in more A-ending codons in mtDNA than in nuclear genome. A new finding summarized below appears to favour THCU.

Zischler *et al.* (1995) discovered a segment (540 bp) of the human mitochondrial D-loop to have been inserted into the nuclear genome, and that the inserted sequence has presumably existed as non-functional DNA. The nucleotide frequencies of the insert for A, C, G, and T are 30.7%, 32.6%, 13.9%, and 22.8%, respectively. The equivalent values for the homologous 540 bp D-loop segment (from LOCUS HUMMTCG in GenBank) are 30.4%, 32.8%, 14.1%, and 22.8%, respectively. If mutations are more biased in favour of A in mtDNA than in nuclear genome, then we should expect a reduction of the proportion of A in the insert, which is not true.

Zischler *et al.* (1995) also sequenced the nuclear DNA sequences flanking the insert. The two flanking regions add up to a total of 385 bp, with the nucleotide frequencies being 41.3%, 18.2%, 13.5%, and 27.0%, respectively, for A, C, G, and T. Thus, the A-content of the non-functional DNA of nuclear origin appears to be in excess rather than in deficiency in comparison with the equivalent values in mitochondrial D-loop. This suggests that mutations in mtDNA is not more biased in favour of A than those in the nuclear genome. In short, the larger proportion of A-ending codons in mtDNA relative to nuclear DNA is not due to mutation bias favouring A in mtDNA.

It is much more difficult to distinguish THCU from the mutation hypothesis regarding the differences in codon usage among mammalian species of different metabolic rates (Tables 3-4). For example, although the proportion of A-ending codons is greater for mouse genes than for rat genes, the proportion of T-ending codons (P_T) also seems to be greater for the mouse genes than for the rat genes. This concurrent increase in both P_A and P_T in animals of higher metabolic rate (i.e., the mouse) is compatible with the mutation hypothesis (Martin 1995) invoking mutation bias favouring A or T in animals of higher metabolic rate (SRM). However, for the nine codon families with both A-ending and T-ending codons, P_A for the mouse genes is significantly larger than P_A for the rat genes ($P = 0.017$, paired T-Test, one-tailed), whereas the difference in P_T between the mouse genes and the rat genes is not significant ($P = 0.507$). This suggests that THCU is a plausible alternative to the mutation hypothesis.

The data of introns (Table 4) are almost entirely compatible with the mutation hypothesis in that a concurrent increase in both A-content and T-content is observed in genes from mammalian species with a high metabolic rate relative to those from mammalian species with a low metabolic rate. The only exception involves comparisons between human and mouse for the β-globin gene. The mouse introns for the β-globin gene show higher A-content and lower T-content than human introns. This is expected under THCU, but not under the mutation hypothesis. However, such a single case should not be taken as a rejection of the mutation hypothesis, which to me remains a plausible hypothesis in many other cases.

I conclude that THCU is a sufficient, and perhaps unique, explanation for the biased codon usage favouring A-ending codons in mammalian mtDNA (Table 1), and the differences in codon usage between the mitochondrial genomes and the nuclear genomes (Table 2). My results further suggest that THCU is a plausible hypothesis in explaining the differences in codon usage in nuclear genomes among mammalian species of different metabolic rates (Table 3-4).

5. DISCUSSION

The prevailing hypothesis on the evolution of codon usage suggests that the pattern of synonymous codon bias is a consequence of adaptation of codon usage to relative availability of tRNA's in the cellular matrix (reviewed by Ikemura 1992). A more relaxed hypothesis invokes the mutual adaptation of codon usage and tRNA availability (Bulmer 1988). According to this second hypothesis, there are three elements in the system determining the evolution of codon usage: mutation bias, tRNA availability, and random genetic drift (Bulmer 1991). Random genetic drift could lead to biased codon usage and unequal availability of different tRNA's in the absence of natural selection. If a synonymous codon that drifts to high frequency happens to be the one recognized by the most abundant tRNA, or if a tRNA that drifts to high abundance happens to be the one that recognizes the most frequently used codon, then these genetic drifts would result in increased translational efficiency and accuracy, and would therefore be favoured by natural selection. This would ultimately result in the most frequently used synonymous codon being recognized by the most abundant tRNA's (Gouy and Gautier 1982; Ikemura 1981; Ikemura 1982; Ikemura 1985; Ikemura 1992; Ikemura and Ozeki 1983). In short, the second hypothesis suggests that mutation bias, tRNA availability and random genetic drift form a self-contained system such that the interaction among the three elements is sufficient to explain the pattern of codon usage.

The results in this paper indicate that this second hypothesis is too restrictive because some features of codon usage, such as the usage of A-ending codons, depends on factors that are not contained in the system of the three elements specified in that hypothesis. Specifically, our optimality model of the transcriptional process predicts that the pattern of synonymous codon usage should depend on the relative concentration of nucleotides in the cellular medium. This is consistent with the findings that the mitochondrial genome has a greater proportion of A-ending codons than the nuclear genome and that the nuclear genome in organisms with a high metabolic rate has a greater proportion of A-ending codons than the nuclear genome in organisms with a low metabolic rate. Thus, a more complete theory of the evolution of codon usage should consider the relative availability of ribonucleotides in the cellular matrix.

One potential misunderstanding concerning THCU and its predictions on biased usage of ATP in transcription is that, because ATP and GTP were used as energy sources in cellular processes, the use of ATP would tend to deplete available energy sources. The benefit of using ATP to enhance the transcription would consequently be offset by the cost of depleting the available energy sources. This argument arises from a misunderstanding that CTP and UTP can come free without spending ATP to synthesize them. It is in fact energetically more efficient to use ATP directly to fill a nucleotide site than to use ATP to synthesize an alternative NTP and then use that alternative NTP to fill in the nucleotide site. In other word, using ATP directly in transcription not only speeds up transcription, it also **conserves** available energy sources.

I should admit here that, although predictions from the model appears consistent with empirical data, the construction of the model itself is not vigorous because of simplifying assumptions. Protein synthesis is a multi-step process including initiation of transcription, elongation of mRNA chain, initiation of translation, and elongation of the peptide chain. By assuming that the rate of transcription rate is limiting, we have reduced the multi-step process to a one-step process, which obviously is a distortion of the reality. However, the recognition that even a very simple model could account for a substantial amount of variation in codon (nucleotide) usage would help to reduce the mystique surrounding the operation of natural selection on the biochemical systems in the living cell.

Chapter 10

Case Study 3

Translational Efficiency and Codon Usage Bias

1. INTRODUCTION

Synonymous codon usage differs among different genomes (Grantham et al. 1981; Grantham et al. 1980; Martin 1995; Moriyama and Hartl 1993; Xia 1996), among different genes within the same genome (Gouy and Gautier 1982; Ikemura 1985; Ikemura 1992; Sharp et al. 1988; Sharp and Li 1987; Sharp and Mosurski 1986), and even among different segments of the same gene (Akashi 1994a). Three hypotheses have been proposed to account for this variation of synonymous codon usage (or various components of the variation): the mutation bias hypothesis (Martin 1995), the transcription-maximization hypothesis (Xia 1996) and translational efficiency hypothesis (Bulmer 1988; Bulmer 1991; Ikemura 1981; Kimura 1983; Kurland 1987a; Kurland 1987b; Robinson 1984; Xia 1998a).

Of these three hypotheses, the translational efficiency hypothesis (hereafter referred to as TEH) is the most general and has received the most empirical support. In verbal forms, the hypothesis states that there is strong selection favouring increased rate of protein synthesis and that a coding strategy that increases the rate of translation initiation and peptide elongation (and consequently increases the rate of protein synthesis) is favoured by natural selection. The hypothesis is favoured by three independent lines of evidence. First, the frequency of codon usage is positively correlated with tRNA availability (Gouy and Gautier 1982; Ikemura 1981; Ikemura 1982; Ikemura 1985; Ikemura 1992; Ikemura and Ozeki 1983). Second, the degree of codon usage bias is related to the level of gene expression, with highly expressed genes exhibiting greater codon bias than lowly expressed genes

(Bennetzen and Hall. 1982; Ikemura 1985; Sharp et al. 1988; Sharp and Devine 1989). Third, mRNA consisting of preferred codons is translated faster than mRNA artificially modified to contain rare codons (Robinson et al. 1984; Sorensen et al. 1989).

Many models of TEH have been presented that can be called either initiation models or elongation models. Initiation models assume that the initiation of translation is rate-limiting (e.g., Bulmer 1991; Liljenström and vonHeijne 1987; Xia 1996), whereas elongation models assume that the elongation of the peptide chain is rate-limiting (Bulmer 1988; Varenne et al. 1984). Empirical data and theoretical considerations suggest that both initiation and elongation are rate-limiting.

The model presented here is strictly a deterministic elongation model, because I think that previous elongation models are not well presented and that expectations are often only vaguely specified. This has resulted in some confusion. For example, Kimura (Kimura 1983) assumed that the translational efficiency is maximized when the proportion of different synonymous codons matches exactly the proportion of isoaccepting tRNAs. The assumption is unwarranted, and the translational efficiency, given the perfect matching, will be shown later to be the same as the presumably less adaptive scenario when different tRNA species are present in equal amount and codon usage drifts freely in any direction.

Another reason for presenting the model is to relate amino acid usage to the availability of tRNA species carrying different amino acids. From an evolutionary point of view, one would intuitively expect an efficient translational machinery to have more tRNA coding for more frequently used amino acids, but this intuition has not been formally established or rejected.

Below I present the elongation model, from which a few specific predictions concerning mutual adaptation between tRNA content and codon usage are derived. Also derived is a relationship between tRNA content and amino acid usage. Empirical data from *Escherichia coli* were used to test the predictions.

2. THE ELONGATION MODEL, ITS PREDICTIONS, AND EMPIRICAL TESTS

Consider the time required to translate a single codon coding for amino acid i (AA_i, i = 1, 2, ..., 20). Designate this codon as SC_{ij} (j = 1, 2, ..., n_i, where n_i is the number of synonymous codons for AA_i). Let r be the rate of aminoacyl-tRNA diffusing to the A site of the ribosome during translation, P_i be the probability that the arriving aminoacyl-tRNA carries AA_i ($\sum_{i=1}^{20} P_i = 1$), p_{ij} be the conditional probability that the aminoacyl-tRNA recognizes the synonymous codon SC_{ij}, given that the tRNA carries AA_i

($\sum_{j=1}^{n_i} p_{ij}=1$ for each given i). Let t_l be the time spent in linking the right amino acid to the elongating protein chain, and t_r be the time spent in rejecting each wrong aminoacyl-tRNA. Now the total time spent in translating SC_{ij} is:

$$T_{ij}=\frac{1}{r\,P_i\,p_{ij}}+t_l+\left(\frac{1}{P_i\,p_{ij}}-1\right)t_r \qquad (10.1)$$

where the first term on the right-hand side of the equation is the time needed for an aminoacyl-tRNA carrying the right amino acid and the right cognate anti-codon to arrive at the A site of the ribosome, and the third term represents time spent in rejecting all the wrong aminoacyl-tRNA prior to the arrival of the right aminoacyl-tRNA. Similar formulation can be found in Varenne et al. (1984) and Bulmer (1988). The total time (T) required to translate L codons (total elongation time) can be shown to be

$$T=\sum_{i=1}^{20}\sum_{j=1}^{n_i} f_{ij}\,T_{ij}=L(t_l-t_r)+\frac{1+r\,t_r}{r}Y \qquad (10.2)$$

where

$$Y=\sum_{i=1}^{20}\frac{1}{P_i}\sum_{j=1}^{n_i}\frac{f_{ij}}{p_{ij}}=\sum_{i=1}^{20}\frac{N_i}{P_i}\sum_{j=1}^{n_i}\frac{Q_{ij}}{p_{ij}}$$

The term f_{ij} is the frequency of synonymous codon j for amino acid i in the mRNA molecule ($\sum_{i=1}^{20}\sum_{j=1}^{n_i} f_{ij}=L$), N_i is the number of codons for amino acid i ($\sum N_i = L$; $\sum_{j=1}^{n_i} f_{ij} = N_i$), and Q_{ij} is the proportion of synonymous codon j for amino acid i in the mRNA molecule. Note that Q_{ij} is a property of the mRNA whereas P_i and p_{ij} are properties of the tRNA pool, with P_i being the proportion of tRNA carrying AA_i, and p_{ij} being the fraction of tRNA that recognizes synonymous codon j among all tRNA species that carry AA_i.

Our objective is to find the condition (i.e., the relationship among Q_{ij}, P_i and p_{ij}) that minimizes T. Because t_l, t_r and L are not dependent on Q_{ij}, P_i and p_{ij}, they are treated as constants. Thus, minimizing T in equation (10.2) is equivalent to minimizing Y. Specifically, we are interested in three relationships. First, given the relative availability of different tRNA (P_i and p_{ij}), find what pattern of codon usage (Q_{ij}) in the mRNA would minimize Y. Second, given the pattern of codon usage (Q_{ij}), find what values for P_i and p_{ij} would minimize Y. Third, given amino acid usage, find the distribution of

P_i that would minimize Y. Intuitively, we would expect frequently used amino acids to correspond to large P_i values, but the exact relationship has not been derived, let alone tested against empirical evidence.

2.1 Adaptation of Codon Usage to tRNA Content

Suppose that an mRNA molecule specifies N residues of the same amino acid with n synonymous codons, and that the associated frequency distribution of synonymous codons is Q_j ($\sum Q_j = 1$). For simplicity, we assume that there are also n types of tRNA species for the amino acid, with each type recognizing just one of the n synonymous codons. The proportion of the n types of tRNA species is p_j ($\sum p_j = 1$). Now we have Y (which is the term to be minimized) below:

$$Y = \frac{N}{P}\sum_{j=1}^{n}\frac{Q_j}{p_j} \tag{10.3}$$

First consider what values Q_j should take when $p_j = 1/n$. One might intuitively think that, to make full use of the equal availability of the n types of tRNA, Q_j should match p_j and should all be equal to $1/n$. This is false. When p_j values are equal, Y is equal to (n*N /P) no matter what value Q_j takes as long as Q_j values sum to 1. Thus, Q_j is a neutral character when p_j values are all equal. I reiterate this point because some confusion has been introduced by Kimura (1983) who wrongly assumed that the highest translational efficiency is achieved when the relative frequencies of synonymous codons exactly match those of the cognate tRNAs.

When p_j values are not equal, then the smallest Y is achieved when the codon recognized by the most abundant tRNA becomes fixed, with the consequent loss of other synonymous codons. To see this more clearly, we re-write equation (10.3) as follows:

$$Y = \frac{N}{P}\,\frac{Q_1\prod_{j=2}^{n}p_j + Q_2\prod_{j=1,j\neq 2}^{n}p_j + Q_3\prod_{j=1,j\neq 3}^{n}p_j + \ldots + Q_n\prod_{j=1,j\neq n}^{n}p_j}{\prod_{j=1}^{n}p_j} \tag{10.4}$$

If p_1 is the largest of all p_j values, then the first Π term (i.e., the one associated with Q_1) on the numerator of equation (10.4) is the smallest of all Π terms. It is therefore obvious that minimization of Y in equation (10.4) requires that Q_1 equal 1 and that all other Q_j values equal zero. This means that whenever the availability of different tRNA species (p_j) for an amino acid is different, the codon usage of this amino acid should evolve towards increasing the frequency of the synonymous codon that is recognized by the most abundant cognate tRNA species. The minimum of Y achievable through adaptation of codon usage to tRNA content is:

$$Y_{\min} = \frac{N}{P\,p_1} = \frac{N}{P\,p_M} \tag{10.5}$$

where p_M designates the most abundant tRNA species for the amino acid. Y_{min} reaches its minimum value when $p_M = 1$, which requires not only the adaptation of codon usage to tRNA content, but also adaptation of tRNA content to extremely biased codon usage.

For the special case with $n = 2$, Y in equation (10.3) can be written as

$$Y = \frac{N}{P}\left(\frac{Q_1}{p_1} + \frac{1-Q_1}{1-p_1}\right) \tag{10.6}$$

The term within the parenthesis is plotted against p_1 and Q_1 (fig. 1). Two conclusions can be drawn. First, When p_j values are all equal to $1/n$ (i.e., when $p_1 = 0.5$ in fig. 1 for $n = 2$), then Q_j can take any value between 0 and 1 without affecting translational efficiency, and Y is relatively small. We will call this condition with equal p_j values as the baseline condition. For unequal p_j values (i.e., for $p_1 \neq 0.5$ in fig. 1), Y values will be larger than that in the baseline condition whenever Q_j values are smaller than p_j values for $p_j > 1/n$ (e.g, when $Q_1 = 0.8$ and $p_1 = 0.9$ in fig. 1) or larger than p_j values for $p_j < 1/n$ (e.g, when $Q_1 = 0.9$ and $p_1 = 0.1$, in fig. 1), in which case the reduction in translational efficiency (i.e., the increase in Y) is outstanding (fig. 1). Y will be the same as that in the baseline condition when Q_j exactly matches p_j (e.g., when $Q_1 = p_1$ in fig. 1). The baseline condition therefore seems to guarantee a relatively small Y value over a wide fluctuation of Q_j values. Y will be smaller than the baseline condition only when Q_j values are larger than p_j values for $p_j > 1/n$ (e.g, when $Q_1 = 0.9$ and $p_1 = 0.8$ in fig. 1) or smaller than p_j values for $p_j < 1/n$ (e.g, when $Q_1 = 0.1$ and $p_1 = 0.2$, in fig. 1).

We have now reached a specific and intuitively appealing prediction, that codon usage bias should be more extreme than the bias in tRNA content. If p_j is larger than $1/n$, then Q_j should be larger than p_j; if p_j is smaller than $1/n$, then Q_j should be smaller than p_j. If this is not the case, then the translational efficiency is lower than that for the baseline condition.

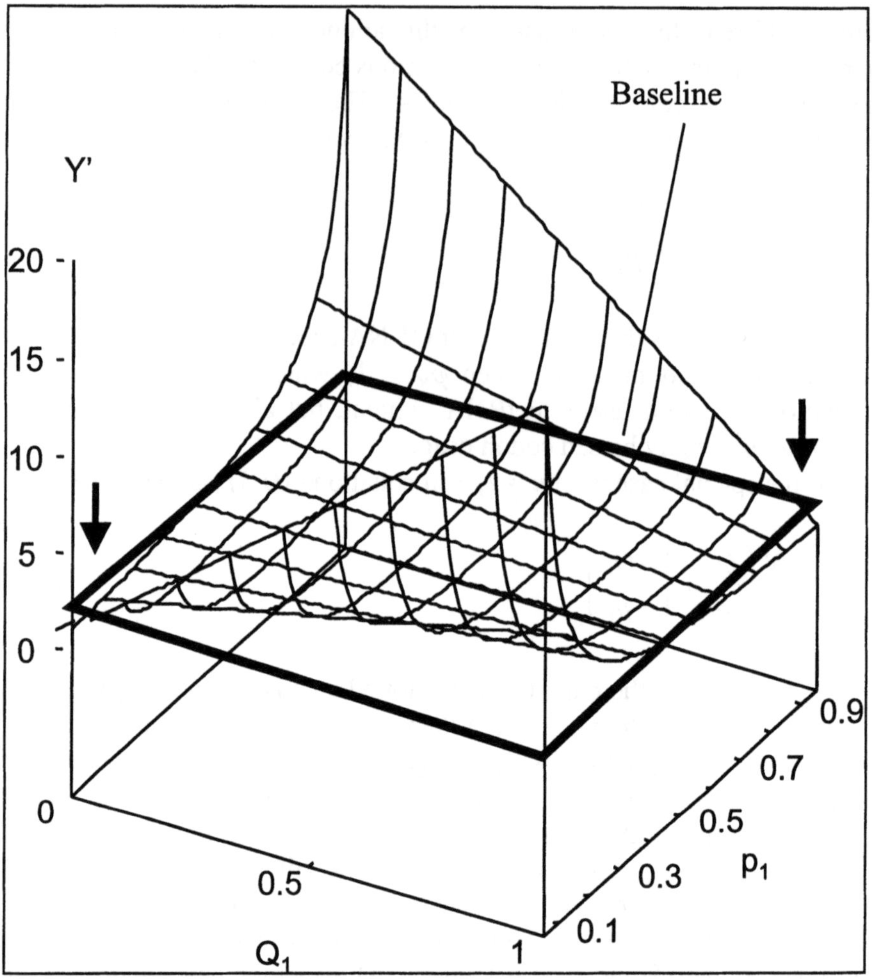

Figure 1. Change in translational time in relation to Q_1 (the proportion of codon 1 in a two-codon family) and p_1 (the proportion of tRNA species recognizing codon 1). Y' is the term within the parenthesis in equation (10.6). The bolded plane perpendicular to Y' represents the "baseline condition" (i.e., the Y' value for $p_1 = 0.5$). The downward arrows designate areas where Y' is smaller than it is in the baseline condition.

An empirical test of this prediction has several requirements. First, we need codon families in which a codon will not be recognized by both the common and the rare tRNA, otherwise Q_j would be impossible to calculate in any meaningful way. Among the 23 codon families (i.e., when we split each of the six-member codon families for Leu, Ser, and Arg into two), only six meet this criterion (Table 1). Secondly, we need codon usage of genes that are highly expressed, otherwise we should not expect any mutual adaptations between tRNA content and codon usage bias. Ikemura (1992) compiled codon usage of presumably highly expressed genes in *E. coli* (five genes), which are used to generate Table 1.

Table 1. Adaptation of codon usage to tRNA content in *E. coli.* ID_{tRNA} – specific tRNA designation; codon – codons recognized by the corresponding tRNA species; N_{tRNA} – number of tRNA genes. tRNA – tRNA content from Table 2 in Ikemura 1992, p_M – the proportion of the most abundant tRNA among all tRNA species carrying the same amino acids. Q_M – the proportion of the codon(s) recognized by the most abundant tRNA among all synonymous codons. Q_M are based on codon frequency data from Table 1 in Ikemura (1992).

				E. coli		
AA	ID_{tRNA}	Codon	N_{tRNA}	tRNA	p_M	Q_M
Gly	3	GGU,GGC	4	1.10	0.815	0.995
	2	GGA,GGG	1	0.15		
	1	GGG	1	0.10		
Ala	1	GCU,GCA,GCG	3	1.00	0.769	0.964
	2	GCC	2	0.30		
Arg	2(1)	CGU,CGC,CGA	4	0.90	0.973	1
	CGG	CGG	1	0.025*		
Ile	1	AUU,AUC	3	1.00	0.952	1
	2	AUA	1	0.05		
Thr	1+3	ACU,ACC	2	0.80	0.800	0.992
	2	ACG	1	0.10		
	4	ACA,ACG	1	0.10		
Gln	2	CAG	2	0.40	0.571	0.954
	1	CAA	2	0.30		

* = the tRNA content is reported as "minor". I used half of the smallest value.

For all three species, the Q_M values are always larger than the p_M values (Table 1). This guarantees that the resulting Y is smaller than that in the baseline condition. The adaptation of codon usage to tRNA content in the highly expressed genes in the three unicellular species is almost perfect (the optimal is when $Q_M = 1$), suggesting that the effect of mutation on codon usage bias must be very weak for these genes. However, if we ignore the expressivity of the genes and pool the codon usage of all genes in the gene bank, then most Q_M values are smaller than the p_M values (data not shown), suggesting that, for most genes, the translational efficiency is lower than that in the baseline condition.

2.2 Adaptation of tRNA to Codon Usage

When Q_j values are fixed (e.g., when codon bias is maintained by mutation bias), the values that p_j should take to minimize Y can be found as follows. We first re-write Y in equation (10.3):

$$Y = \frac{N}{P}\left(\frac{Q_1}{1 - \sum_{j=2}^{n} p_j} + \sum_{j=2}^{n} \frac{Q_j}{p_j} \right) \tag{10.7}$$

The condition that minimizes Y is found by taking partial derivatives of Y with respect to p_j, and setting the partial derivatives to zero. This yields:

$$\frac{Q_1}{p_1^2} = \frac{Q_2}{p_2^2} = \ldots = \frac{Q_3}{p_n^2} \tag{10.8}$$

Expressed in another way, the condition implies:

$$\frac{p_j}{p_k} = \sqrt{\frac{Q_j}{Q_k}} \tag{10.9}$$

i.e., the bias in tRNA availability for an amino acid should not be as dramatic as that in codon usage. In other words, selection driving tRNA adaptation to codon usage guarantees that tRNA bias will not be as extreme as codon bias. Results similar to equation (10.9) have been derived before (Bulmer 1988).

The relationship between p and Q in equation (10.9) can also be written as $p = a\sqrt{Q}$, where a is a constant. Ikemura (1992) plotted an equivalent measure of Q versus an equivalent measure of p (fig. 3 in Ikemura 1992) for a highly expressed gene in *E. coli* (*groEL*), and the result confirmed the predicted quadratic relationship between p and Q.

We should now note that the baseline condition depicted in fig. 1 is not stable because, with all p_j values equal to $1/n$, Q_j values can drift to any value without affecting translational efficiency (equation (10.3) and fig. 1. When Q_j values differ from $1/n$, there will then be selection favouring adaptation of tRNA content to codon usage (equation (10.9)), which would drive p_j values away from $1/n$. Note that this selection pressure will not drive p_j values more extreme than Q_j values (equation (10.9)), otherwise the selection would result in a less efficient translational machinery. The resulting unequal p_j

values, in turn, creates selection pressure for codon usage adaptation (equation (10.5)).

2.3 Evolution of tRNA in Response to Amino Acid Usage

To my knowledge, none of the TEH models linked tRNA availability to amino acid usage. The 20 amino acids are not used equally in proteins, and we intuitively would expect those frequently used amino acids to be carried by more tRNA than those rarely used amino acids. To better visualize the effect of amino acid usage on P_i, which is the proportion of tRNA species carrying amino acid i in the total tRNA pool, we write Y in equation (10.2) in the expanded form:

$$Y = \frac{N_1}{P_1}\left(\frac{Q_{1,1}}{p_{1,1}} + \frac{Q_{1,2}}{p_{1,2}} + \ldots + \frac{Q_{1,n_1}}{p_{1,n_1}}\right)$$

$$+ \frac{N_2}{P_2}\left(\frac{Q_{2,1}}{p_{2,1}} + \frac{Q_{2,2}}{p_{2,2}} + \ldots + \frac{Q_{2,n_2}}{p_{2,n_2}}\right) + \ldots \qquad (10.10)$$

$$+ \frac{N_{20}}{P_{20}}\left(\frac{Q_{20,1}}{p_{20,2}} + \frac{Q_{20,2}}{p_{20,2}} + \ldots + \frac{Q_{20,n_{20}}}{p_{20,n_{20}}}\right)$$

where N_i is the total number of codons for amino acid i. When codon usage is perfectly adapted to tRNA availability for each amino acid, which is approximately true based on empirical data in Table 1, Y becomes:

$$Y = \frac{N_1}{P_1\, p_{M_1}} + \frac{N_2}{P_2\, p_{M_2}} + \ldots + \frac{N_{20}}{P_{20}\, p_{M_{20}}} \qquad (10.11)$$

according to equation (10.5). The minimization of Y requires

$$\frac{P_i}{\sqrt{N_i p_{M_j}}} = \frac{P_j}{\sqrt{N_j p_{M_i}}} \qquad (10.12)$$

where P_i and P_j designate the proportion of tRNA carrying amino acids i and j, respectively; and N_i and N_j are the number of amino acids i and j, respectively. When tRNA concentration for each amino acid is well adapted to codon usage, all p_M values approach 1 and become nearly equal, so that equation (10.12) becomes:

$$\frac{P_i}{\sqrt{N_i}} = \frac{P_j}{\sqrt{N_j}} \text{ or } P = a\sqrt{N} \tag{10.13}$$

Empirical data for testing the above prediction is readily available. The P_i values can be derived from data in Table 2 in Ikemura (1992) for *E. coli*. Ikemura (1992) also compiled the codon usage of 937 *E. coli* genes, from which one can derive N_i values in equation (10.13). The 20 pairs of P_i and $\sqrt{N_i}$ values are plotted on fig. 2 for *E. coli*. The fit is quite remarkable.

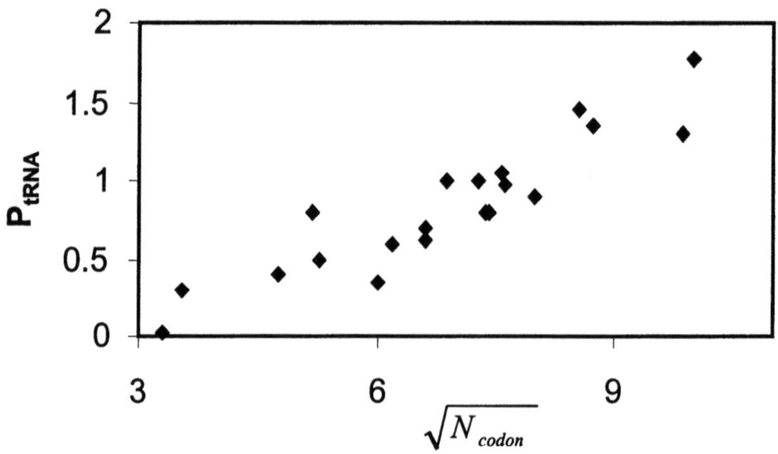

Figure 2. The availability of tRNA carrying a certain amino acid increases linearly with the square-root of the frequency of the amino acid in *E. coli*. The 20 N_i values are from Table 1 in Ikemura (1992). Corresponding P_i values are from Table 2 in Ikemura (1992). N_i values were presented as the number in 1000.

Such a seemingly straightforward interpretation, however, has a major difficulty. The argument requires that all p_M values be either approximately one (which should hold only for highly expressed genes), or approximately equal (which we have no reason to expect), so as to cancel each other out. Only a few loci are deemed highly expressed, yet 937 loci from *E. coli* were used for fig. 2. Why should lowly expressed genes contribute to the linear relationship? The simplifying assumption, that $p_M \approx 1$, seems unjustified. It is therefore necessary to work out the relationship between P and N when the assumption of $p_M \approx 1$ does not hold.

I propose the following equation, which is more general than equation (10.13) and does not require $p_M \approx 1$, to describe the relationship between P and N:

$$P = aN^b \tag{10.14}$$

If the parameter b is shown to be = 1/2, then equation (10.14) is reduced to equation (10.13). Given equation (10.14), we have

$$\frac{P_i}{P_j} = \frac{aN_i^b}{aN_j^b} = \frac{\sqrt{N_i}}{\sqrt{N_j}} \cdot \frac{\sqrt{p_{M_j}}}{\sqrt{p_{M_i}}} \tag{10.15}$$

After some algebraic manipulation, we obtain

$$P = aN^{\frac{1+Z}{2}} \tag{10.16}$$

where

$$Z = \frac{\ln \dfrac{p_{M_j}}{p_{M_i}}}{\ln \dfrac{N_i}{N_j}} \tag{10.17}$$

As expected, the relationship between P and N depends on the magnitude of Z, which in turn depends on the relationship between p_M and N. If p_M is independent of N and approaches 1, then $Z = 0$, and $P = aN^{1/2}$, which is equation (10.13). If p_M and N are positively correlated, then $Z < 0$. If Z lies within (-1, 0), then P will increase with N at a decreasing rate. If $Z = -1$, then there will be no relationship between P and N, which we know to be false from fig. 2. If $Z < -1$, then P will decrease with N at a decreasing rate, which we also know to be false from fig. 2. If p_M and N are negatively correlated, then $Z > 0$. If Z is between 0 and 1, then P will increase with N at a decreasing rate. If $Z = 1$, then P will increase linearly with N, rather than with the square-root of N as predicted from equation (10.13). If $Z > 1$, then P will increase with N at an increasing rate.

There seems to be a slightly negative relationship between p_M and N for data from the two prokaryotic species (fig. 3), which is not statistically significant. Based on the relationship between p_M and N for the two prokaryotic species, we expect Z in equations (10.16) and (10.17) to be slightly larger than 0. Consequently, the coefficient b in $P = aN^b$ should be slightly larger than 1/2. The b values that provide the best fit to the data points in fig. 2 are 1.10 (i.e., Z = 1.20). Equation (10.14), however, does not fit the empirical data significantly better than equation (10.13).

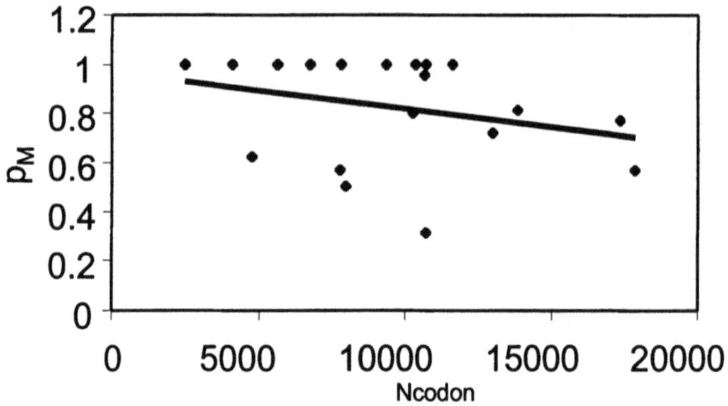

Figure 3. The slightly negative relationship between p_M (the proportion of the most abundant tRNA among all tRNA species carrying the same amino acids) and N_{codon} (the number of codons for each amino acid). *E. coli* data.

2.4 Translational Efficiency and Translational Accuracy

Translational accuracy has recently been suggested to be an important factor related to codon usage bias (Akashi 1994b; Bulmer 1991). This proposal received empirical substantiation from a study of protein-coding genes in Drosophila that revealed differences in codon usage among different regions of the same gene. For example, gene regions of greater amino acid conservation tend to exhibit more dramatic codon usage bias than do regions of lower amino acid conservation (Akashi 1994b).

Translational efficiency and translational accuracy are inextricably coupled in their effect on codon usage bias. To reduce translational error, one needs to reduce the number of wrong aminoacyl-tRNA species that have to be rejected before the arrival of the right aminoacyl-tRNA. Equation (10.1) shows this number to be

$$N_{wrong} = \left(\frac{1}{P_i \, p_{ij}} - 1 \right) \tag{10.18}$$

for each codon translated. To translate an mRNA with L codons, the total number of wrong aminoacyl-tRNA species that translational machinery needs to reject is

$$N_{L.wrong} = Y - L \tag{10.19}$$

where Y is exactly the same as the Y in equation (10.2). To minimize the number of translational errors, we minimize Y, which leads to exactly the same predictions that we have already attributed to TEH. The rationale for separating the effect of maximizing translational accuracy on codon usage bias from that of maximizing translational efficiency is discussed latter.

3. DISCUSSION

3.1 Validity of the Model

Protein synthesis is a multi-step process including initiation of transcription, elongation of mRNA chain, initiation of translation, and elongation of the peptide chain. Opinions differ concerning which step might be rate-limiting. Xia (1996) argued that the rate of protein synthesis depends much on the rate of initiation of translation. He reasoned that the rate of initiation depends on the encountering rate between ribosomes and mRNA, which in turn depends on the concentration of ribosomes and mRNA. Thus, patterns of codon usage that increase transcriptional efficiency should increase mRNA concentration, which in turn would increase the initiation rate and the rate of protein synthesis. He presented a model predicting that the most frequently used ribonucleotide at the third codon sites in mRNA molecules should be the same as the most abundant ribonucleotide in the cellular matrix where mRNA is transcribed. This prediction is supported by several lines of evidence. That the initiation step is rate-limiting has also been suggested by other studies (Bulmer 1991; Liljenström and vonHeijne 1987).

While not denying the possibility that initiation of translation may be rate-limiting, the model presented here explicitly assumes that the elongation of the peptide chain is rate-limiting. There is a substantial amount of empirical evidence supporting this assumption (Bonekamp et al. 1985; Bonekamp and Jensen 1988; Pedersen 1984; Williams et al. 1988). In particular, mRNA consisting of preferred codons is translated faster than mRNA artificially modified to contain rare codons (Robinson et al. 1984; Sorensen et al. 1989). That elongation is a rate-limiting process has also been suggested on the basis of theoretical considerations (Liljenström and vonHeijne 1987).

Bulmer (1991), however, argued that initiation rather than elongation is rate-limiting. He reasoned that, for elongation to be rate-limiting, there should be so many ribosomes that would bind to all free mRNA molecules as soon as the latter become available for binding. Since ribosomes form the largest part of the protein translational machinery (and therefore likely to be costly and time-consuming to make), it would be inefficient to saturate the system with them. He summarized empirical evidence that seems to suggest that ribosomes are far from saturating the system. For example, there are an average of 225 bases per ribosome in a polysome (Ingraham et al. 1983), and each ribosome covers only about 30 bases (Kozak 1983). This Bulmer (1991) interprets to mean that it is very rare for more than one ribosome to compete for the free binding site of the mRNA. Thus, there is no need for the ribosome to travel down the length of the mRNA in a hurry (i.e., there is little benefit associated with more efficient elongation.

There are two weaknesses in such arguments. First, Kozak's (1983) study does not necessarily mean that a ribosome needs clear only 30 bases to free the initiation site for the binding of the next ribosome. Secondly, even if the ribosome needs to move only 30 bases to free the initiation site, there is still some probability for more than one ribosomes to arrive at the free initiation site. Only one of the arriving ribosomes would have a chance to bind to the initiation site, while the rest would have to be turned away. Increased elongation rate would reduce the occurrence of such events.

In addition to the assumption that elongation is rate-limiting, the model also assumes that either r (i.e., the rate of aminoacyl-tRNA diffusing to the A site of the ribosome during translation) is not extremely large, or t_r (i.e., the time spent in rejecting each wrong aminoacyl-tRNA) is not negligibly small. These seem to be reasonable assumptions, although t_r might indeed be very small (Bilgin et al. 1988).

3.2 Relative Importance of Translational Efficiency and Accuracy on Codon Usage Bias

Although the model of maximizing translational accuracy and that of maximizing translational efficiency produce the same set of predictions, it is still possible to separate the effect of maximizing translational accuracy on codon usage bias from that of maximizing translational efficiency. For example, a protein gene could have arginine codons in different domains of different functional importance. Being in the same protein gene, these arginine codons are subject to the same selection pressure exerted by maximizing translational efficiency, and consequently should have the same codon usage bias according to the model of maximizing translational efficiency. However, those arginine codons located in the functionally important domains are subject to greater selection pressure exerted by maximizing translational accuracy than those located in the functionally unimportant domains. Consequently, the former codons will be more biased towards using the optimal codon than the latter. Some preliminary findings along this line of reasoning have already been reported (Akashi 1994a).

The reasoning above leaves one question unanswered. Why is it necessary to invoke translational efficiency to account for codon usage bias? Can't we attribute all the codon usage bias to the effect of maximizing translational accuracy and forget about translational efficiency? The answer is that the effect of maximizing translational accuracy is insufficient to account for the observed codon usage bias. For example, highly expressed genes exhibit greater codon bias than lowly expressed genes, but the former are not necessarily more conservative than the latter (greater conservativeness presumably implies greater demand for accuracy). We can rank protein genes according to their conservativeness, or rank them according to their expressivity, and find out which ranking explains codon usage bias better. Preliminary results (unpublished) suggest that the expressivity is the more important of the two.

It should be noted that the within-gene variation in codon usage bias found in Drosophila (Akashi 1994a) does not seem to be general. For example, it is not observed in *E. coli* and *S. typhimurium* (Hartl et al. 1994a). More empirical studies are needed to assess the effect of maximizing translational accuracy on codon usage bias.

3.3 How Optimized Are the Translational Machinery?

From our results, we can say that codon usage in those highly expressed genes are almost as optimal as possible, with the Q_M values larger than p_M values and almost equal to 1. However, for the majority of genes, the Q_M values are smaller than p_M (data not shown), which implies that the translational efficiency for the majority of the genes is less than in the seemingly less adaptive scenario when different tRNA species are present in equal amounts and codon usage drifts freely in any direction.

We should note that selection for codon adaptation to tRNA content operates on individual genes, whereas selection for the adaptation of tRNA content to codon usage operates at the genome level. Thus, although equation (10.5) suggests that the optimal condition is when both the most abundant tRNA and its cognate codon become fixed, equation (10.9) shows that selection for tRNA adaptation to codon usage will always lag behind codon usage bias.

The most remarkable feature from the model is the prediction relating amino acid usage (N_i) to tRNA content (P_i), which is strongly supported by empirical evidence (fig. 2). A more extensive study is underway to confirm the generality of the relationship.

Chapter 11

Evolution of Amino Acid Usage

1. INTRODUCTION

Suppose you know nothing more than the fact that proteins are made of 20 amino acids. Now if someone asks you about the amino acid usage of one particular protein, say cytochrome-b in human mitochondria, what would be your response? In the lack of any further information, the best guess is simply that each amino acid should account for 5% (= 1/20) of the total number of amino acids in the protein.

But this guess of 5% is a rather uneducated guess. As you have already learned quite a bit about the codons and genetic code, it might have occurred to you that some amino acids are coded by many codons while some others are coded by few. Take the universal genetic code for example, the amino acid methionine is coded by a single codon, i.e., AUG, whereas amino acids arginine and serine are coded by six codons. It therefore seems reasonable to predict that arginine and serine should be used more frequently than methionine. In the extreme (although fictitious) case when one amino acid is coded by all 61 sense codons, then that amino acid would account for 100% of all amino acids used in the protein and the rest of the 19 amino acid would all have a frequency of zero. In short, we are predicting a positive correlation between the frequency of amino acids ($Freq_{AA.i}$, where i = 1, 2, ..., 20) with the number of codons per amino acid ($N_{cod.i}$, where i = 1, 2, ..., 20). Because $Freq_{AA.i}$ would be zero if $N_{cod.i}$ is zero, the relationship between $Freq_{AA.i}$ and $N_{cod.i}$ may be represented simply as $Freq_{AA.i} = \alpha N_{cod.i}$. You will have a chance to empirically test this prediction and to find out which amino acids deviate from this rule.

The prediction above misses some important information. For example, lycine is coded by only two codons, AAA and AAG, in the universal genetic code, and we consequently expect the amino acid to be used rarely relative to those amino acids coded by four or more codons. However, mutation pressure might favour A and G in some particular genomes, and we may consequently find AAA and AAG codons quite frequent. This would imply frequent usage of lycine in spite of the fact that the amino acid is coded by only two codons. In contrast, leucine and serine may be used rarely even though each of them is coded by six codons, simply because these codons are CT-rich.

Predictions in published scientific journals are typically less straightforward as those shown above. For example, Xia and Li (1998) postulated that typical amino acids should be used more frequently than idiosyncratic amino acids. This prediction may not make sense at first sight. However, once you know what they mean by "typical" and "idiosyncratic", then the prediction becomes quite comprehensible, or even quite natural.

Some amino acids, such as Leu, are "typical" in that they have a number of similar alternative amino acids which they can mutate to through a single nucleotide substitution, which is often called single-step nonsynonymous codon mutation, or SSNCM. Note that "similar" amino acids mean amino acids that are similar in physico-chemical properties. When a SSNCM occurs at the amino acid site occupied by these typical amino acids, there is a large probability that the replacement amino acid is similar in physico-chemical properties to the original. Such changes typically have very minor effects on the normal function of the protein. In contrast, some other amino acids, such as Arg, are idiosyncratic in that they have few similar alternative amino acids. For such amino acids, almost any SSNCM is a mutation of a large effect and may disrupt the function of the protein. If a protein-coding gene contains a large number of typical amino acids and few idiosyncratic amino acids, then the effect of SSNCM would be reduced. In other words, the protein would be more reliable, and the gene coding the protein will therefore be favoured by natural selection against those that use a lot of idiosyncratic amino acids. This leads to the prediction that typical amino acids should be used more frequently than idiosyncratic amino acids. You will also have a chance to test this prediction using DAMBE.

In contrast to the predictions mentioned above that are rather general, there are scientific predictions (or hypotheses) that are very specific. One specific prediction that can derived based on the same reasoning as above goes like this. Suppose that certain sites in a protein require a polar amino acid to maintain the normal function. Consider Glu and Asp, both of which are polar. When a SSNCM occurs, Asp has a higher probability (0.56), but Glu a lower probability (0.44), of being replaced by another polar amino acid (Epstein 1967). If the protein gene uses more Asp codons and fewer Glu codons for such sites, the function of the protein would less likely be disrupted by mutation. It is therefore beneficial for the fictitious protein gene to use more Asp codons but fewer Glu codons. Are you interested in testing this prediction by using DAMBE?

There are many predictions that could be proposed concerning amino acid usage, but we will stop temporarily here to avoid the dreadful consequence of exhausting your brain and diminishing your interest. Let us now develop a good appreciation of biased amino acid usage.

2. AMINO ACID USAGE BIAS

Start DAMBE if you have not yet done so. Open a file containing either amino acid sequences or protein-coding nucleotide sequences. In the latter case, click **Sequence|Work on amino acid sequences** to translate the nucleotide sequences into amino acid sequences. If you do not have your own sequences, then just open the **virus.fas** file which contains protein-coding nucleotide sequences for the hemaglutinin gene from Influenza A viruses infecting mammalian species. Translate the sequences into amino acid sequences. Note that if you choose, instead of the universal genetic code, any of the other 11 genetic codes, then the translation will be all wrong.

You will be prompted as to where to start translation if the sequences are already aligned, as is the case for **virus.fas**. This is necessary because your sequences may begin with a an incomplete codon. For example, the first codon in a complete protein gene is typically AUG, but your sequences may miss the first two nucleotides. Consequently, your sequences, written in codon form, will look like this: G CCC GGA .... In this case you should translate from the second nucleotide, i.e., beginning with the first complete codon CCC, because the first nucleotide, G, is not translatable. If your sequence is like this: UG CCC GGA ... ..., then you should translate from the third nucleotide. If you have translated wrong, click **Sequence|Work on codon sequences** to revert back to the original nucleotide sequences and re-do the translation with a different starting site.

If your sequences are not aligned, then it is possible that one sequence will be AUG CCC GGA ... ..., while another may be G CCC GGA ... ..., and the third UG CCC GGA ... ..., and so on. DAMBE will not ask you where to start translation, but instead will try to identify where to start. For each sequence, DAMBE will carry out translations three times, starting with first, second or third nucleotide position, respectively. Note that there are only these three possibilities. The translation that have the fewest "*" (for stop codon) in the translated sequence will be taken as the best one. A wrong translation typically has many "*" embedded in it. An ideal translated sequence should contain just one terminating "*" at the end.

Now that you have got amino acid sequences in the DAMBE buffer, click **Seq. Analylsis|Amino acid frequency**. The familiar dialog box appears for you to select which sequences you wish to apply the analysis. Just click the **Add All** button to apply the analysis to all sequences. Click the **Done** button. You will be asked if you wish to plot the frequency of amino acids ($Freq_{AA}$) versus the number of codons per amino acids (N_{cod}). Recall that we have predicted a positive correlation between $Freq_{AA}$ and N_{cod}. So click **Yes** to get the graphic output (fig. 1). Is our prediction fulfilled?

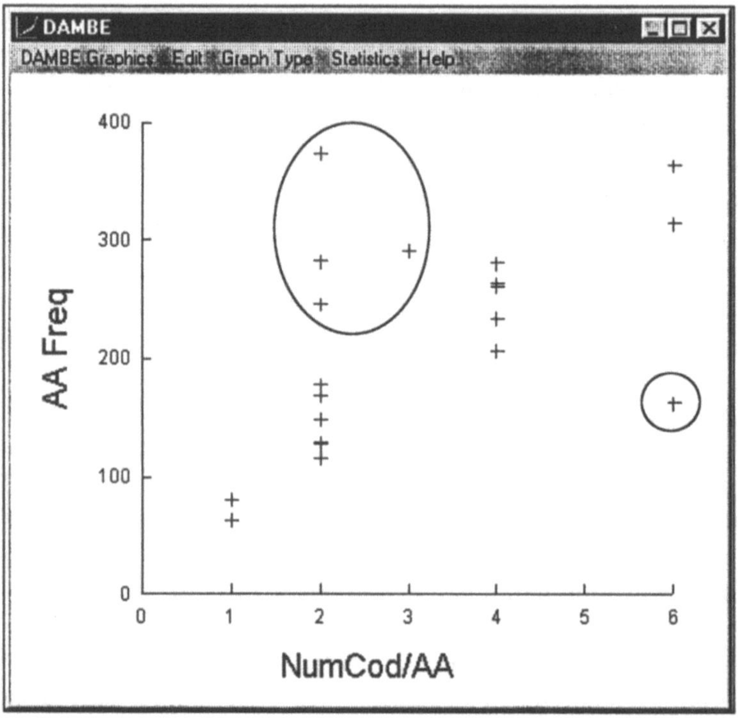

Figure 1. The amino acid frequency is positively correlated with the number of codons per amino acid.

We first note a positive correlation between $Freq_{AA}$ and N_{cod}, which partially confirms our prediction. Second, if we do stick to our prediction and draw a straight regression line across the (0, 0) origin, we see a number of points deviating substantially from the line. In particular, four amino acids (asparagine, glutamine, isoleucine and lycine) have their $Freq_{AA}$ values much above the line whereas one amino acid (arginine) has its FreqAA much below the line. We will come back latter to see if we can offer a satisfactory explanation for the five amino acids that have their $Freq_{AA}$ deviating so much from the prediction.

In addition to the graphic output, other results are saved in a text file and will also be displayed in DAMBE's display window. The output is in two parts. The first (not shown) is a codon-usage table for each of the sequences, and the second is a summary table after pooling all data together. Only the summary output is shown below (and plotted in fig. 1):

```
Pooled output:
AA      Number  Percent  Codons/AA     Pred    Resid
=====================================================
Ala        207     4.82          4   248.43   -41.43
Arg        163     3.80          6   372.64  -209.64
Asn        374     8.71          2   124.21   249.79
Asp        169     3.94          2   124.21    44.79
Cys        116     2.70          2   124.21    -8.21
Gln        246     5.73          2   124.21   121.79
Glu        129     3.00          2   124.21     4.79
Gly        281     6.54          4   248.43    32.57
His        128     2.98          2   124.21     3.79
Ile        291     6.78          3   186.32   104.68
Leu        315     7.34          6   372.64   -57.64
Lys        282     6.57          2   124.21   157.79
Met         63     1.47          1    62.11     0.89
Phe        149     3.47          2   124.21    24.79
Pro        234     5.45          4   248.43   -14.43
Ser        364     8.48          6   372.64    -8.64
Thr        263     6.12          4   248.43    14.57
Trp         81     1.89          1    62.11    18.89
Tyr        178     4.15          2   124.21    53.79
Val        261     6.08          4   248.43    12.57
=====================================================
Sum     4,294

Testing relationship between the number of amino acid
and the number of codons per amino acid:
-------------------------------------------
Pearson R        0.5452
Intercept      121.3432
Slope           30.6088
SigmaB          11.0927
T:               2.7594
DF:             18
```

```
Prob:          0.0032 (One-tailed test)
```

The output above is in two parts. The first part shows mainly a table of amino acid frequencies, designated hereafter as $Freq_{AA}$. We note that the 20 amino acids are not used equally. Although this is not really surprising to those of us who have already compiled amino acid frequencies for a large number of proteins, very few of us are well versed to explain the variation in $Freq_{AA}$. The fourth column shows the number of synonymous codons per amino acid, referred to as NumCod hereafter. Note that NumCod values depend on what genetic code you use. The output above is for the elongation factor-1α from the nuclear genome of invertebrate species. Note that there is only one codon coding for methionine (Met). The value would be two if the sequences are protein-coding genes from mammalian mitochondria. DAMBE implemented 12 different genetic codes.

We have predicted previously that FreqAA should be positively correlated with NumCod. The second part of the output tests whether this prediction is true by calculating the Pearson correlation coefficient and doing a simple linear regression of FreqAA on NumCod. Because we have already predicted a positive correlation between the two, the test is one-tailed, and is marginally significant for most of the genes that I have encountered. The regression typically would account for about 20% of the total variation in amino acid frequencies. In our particular case, the regression accounted for $R^2 \approx 30\%$ of the total variation in amino acid frequency, i.e., the total variation in the column headed by **Number**.

In the unlikely case that you do not know, total variation of a variable in statistics refers to the sum of squared deviations of individual values from the grand mean. In our case, there are 20 individual values, and the mean is simply the length of the peptide divided by 20. If we know nothing more than the mean, then our best predictor of amino acid usage is simply the mean. In this case the total variation is a measure of our ignorance.

The ignorance about FreqAA is reduced significantly when we know NumCod and have established the fact that FreqAA and NumCod are positively correlated. We can incorporate NumCod in predicting the value of FreqAA. The predicted value is also shown in the output under the heading "Pred.", and the residual, which is the observed FreqAA minus the predicted FreqAA, is shown in the column headed by "Resid.". The sum of squared residuals is the remaining variation in FreqAA that we cannot explain, and is now the new measure of our ignorance about FreqAA.

A visual display of the relationship between FreqAA and NumCod shows one point that is particularly off mark. The point is for arginine, which is coded by six codons and is expected to have a large FreqAA value. Note that the regression line is suppose to go through the (0,0) point because FreqAA is necessarily zero if NumCod/AA is zero. We note that FreqAA increases roughly with NumCod linearly, which is what we have expected. But why does arginine have such a small FreqAA value?

Do you have an explanation for this pattern? Interestingly, nobody in the whole world has a ready answer for the reduced amino acid usage for arginine. If you can think of a good answer, then you are right at the frontier of molecular evolution.

One may suggest that it is likely that one family of codons may become evolutionarily disadvantageous when used to code for one particular amino acid, and natural selection would therefore favour the evolution of alternative codons for the same amino acid. For example, it may be deleterious to code the amino acid arginine using CGN codons, and natural selection would therefore favour the evolution of alternative codons such as AGR for the same amino acid. This hypothesis will hereafter be referred to as the codon-switching hypothesis. Can we find some corroborative evidence for this hypothesis?

If the hypothesis is correct, then we should expect very biased codon usage. We should expect very few CGN codons but relatively abundant AGR codons. Now use DAMBE to compile a codon usage table to see if this expectation is correct. Click **Sequences|Work on codon sequences** to restore the nucleotide sequences. Then click **Seq. Analysis|Codon frequency.** Based on the DAMBE output, can you make a judgement on whether the expectation is supported?

The results are consistent with the expectation. Of 30 amino acids in the peptide sequences, 20 were coded by the two AGR codons and only 10 were coded by the four CGN codons.

But why is it not a good idea for the five mammalian species to use CGN codons to code arginine? If we cannot answer this question, then the codon-switching hypothesis remains speculative.

One possible explanation to the observation that CGN codons are rarely used invokes DNA methylation. DNA methylation is a ubiquitous biochemical process observed in both prokaryotes (Noyer-Weidner and Trautner 1993) and eukaryotes (Antequera and Bird 1993). In vertebrates, DNA methylation mainly involves the methylation of C in the CpG dinucleotide, which greatly elevates the mutation rate of C to T through spontaneous deamination of the resultant 5-methylcytosine (Barker et al. 1984; Cooper and Krawczak 1989; Cooper and Krawczak 1990; Cooper and Schmidtke 1984; Cooper and Youssoufian 1988; Ehrlich 1986; Ehrlich et al. 1990; Rideout et al. 1990; Schaaper et al. 1986; Sved and Bird 1990; Wiebauer et al. 1993). This is the mutation effect mediated by methylation that reduces the number of CGN codons in vertebrate genomes.

The elevated mutation pressure implies that a gene with many CpG dinucleotides would be unreliable, especially when the dinucleotide occupies the first and second (CGN) or the second and third (NCG) positions of a codon because a mutation from C in the CpG dinucleotide to T is always nonsynonymous in such cases. Thus, we expect purifying selection to minimize the usage of codons with an embedded CpG dinucleotide. This is the selection effect mediated by methylation that would also reduce the usage of CGN codons.

We are now in a good position to explain why the FreqAA value is so small for arginine (fig. 1). The answer is simply that arginine really is not coded by six codons. It is mainly coded by only two AGR codons, with a small contribution from the four CGN codons. In short, the ancestral codons for arginine may be CGN, but CGN codons become unreliable because of the high mutation rate of CGN codons to TGN codons. Natural selection therefore would favour the coding of arginine by alternative codons, which resulted in arginine being coded also by AGR codons. Because AGR codons do not have the problem of high mutation rate of CGN codons, they become the favoured codons for arginine. Therefore, the NumCod value for arginine should really be just two codons, AGA and AGG, or slightly more than two because CGN codons still code a small fraction of arginine. Given a NumCod value of slightly larger than two, it is not all that extraordinary for arginine to have a small FreqAA value.

So far the codon-switching hypothesis seems to be a story well told. However, it is important to realize that a story well told is not a theory well established. In fact, the reasoning above embedded many jumps of faith. The hypothesis is presented mainly for you to identify its weak points as an exercise to foster critical thinking. Can you present a critical test of the codon-switching hypothesis?

Let us derive one more prediction based on the hypothesis. If the low use of CGN codons in the mammalian species is due to DNA methylation, then we expect a relatively high usage of CGN codons in invertebrate genomes because DNA methylation is week in invertebrate species. We therefore expect arginine to be used more frequently in the invertebrate genomes than in the mammalian species. Can you use DAMBE to test this expectation?

Look at fig. 1 once more. There are four points that are way above the line, and we have not offered any explanation for why the points should be where they are. These points are for asparagine, glutamine, isoleucine and lycine, with the highest point being asparagine. I have searched thoroughly through literature hoping to find an existing answer for you but have failed. If you could find the reason for why asparagine should be used so very frequently, then that is new discovery in science and new knowledge for the human kind.

Chapter 12

Pattern of Nucleotide Substitutions

The rate ratio parameters in substitution models

1. INTRODUCTION

In previous chapters we have learned that a substitution model has two categories of parameters, the frequency parameters and the rate ratio parameters. We have also learned much about the variation in frequency parameters and the factors that affect the frequency parameters. This chapter and the next two chapters will cover fundamentals of the rate ratio parameters. Understanding these chapters will help you better understand the chapter dealing with genetic distances and phylogenetic analysis in later chapters.

The DNA in an organism is like a very long book specifying the development and life cycle of the organism. Every time when a cell divides, the whole "book" is copied, and some copying errors (i.e., mutations) get incorporated in the new "book". If the mutations occur in a somatic cell, then the new "book" is destroyed when the organism dies. However, when the mutations occur in the germ line, then the new "book" has a chance to be passed on to the next generation. The propensity of the "book" being passed on to the next generation is defined in population genetics as the fitness, which obviously depends on the constituent genes in the genome. Whether the "book" carrying the new mutation will be passed on to the next generation again is mainly determined by natural selection and random genetic drift. When a book carrying one particular mutation out-competes all the alternative books carrying other forms of the gene at the same locus, we say that the gene is fixed, and the replacement event is then called a substitution.

What kind of substitutions is the most likely to happen? At the nucleotide level, which is the most fundamental level for any mutation, there are 12 possible changes, with four being transitional changes and eight being transversional changes (Table. 1). Are these 12 different kind of genetic changes equally likely to happen? Or do they all have different substitution rates?

Table 1. Twelve types of nucleotide substitutions which might differ from each other in the probability of occurrence. The rate parameters are designated by a_i.

	To			
From	A	G	C	T
A	$1 - a_1 - a_2 - a_3$	a_1	a_2	a_3
G	a_7	$1 - a_4 - a_5 - a_7$	a_4	a_5
C	a_8	a_9	$1 - a_6 - a_8 - a_9$	a_6
T	a_{10}	a_{11}	a_{12}	$1 - a_{10} - a_{11} - a_{12}$

There are two categories of variables that affect the number of observed substitution event. The first category is the nucleotide frequencies. For example, if DNA sequences are extremely GC-rich, then obviously most observed substitutions will be G↔C. In other words, the observed substitution rate depends on nucleotide frequencies. A nucleotide having a higher frequency than others is more likely to be involved in nucleotide substitutions, everything else being equal. Nucleotide frequencies are referred to as frequency parameters in a substitution model and represent the first category of variables affecting the observed substitution rate. The frequency parameters are generally represented as π_A, π_C, π_G, and π_T. You have already been introduced to the frequency parameters in the chapter dealing with nucleotide frequencies. The amino acid frequencies and codon frequencies that we have compiled from empirical data in previous chapters represent frequency parameters for amino acid-based and codon-based substitution models, respectively.

The second category of variables that affect the substitution rate are typically represented by Greek letters α, β, γ, δ, etc., or simply a_1, a_2, a_3, … a_{12}. If, on average, a C changes to a T in 10,000 years, whereas an A changes to a G in 30,000 years, then $a_{C \to T}$ is three times greater than $a_{A \to G}$. In other words, if we set $a_{A \to G}$ to 1, then $a_{C \to T}$ is 3. If nucleotide frequencies are equal, we would expect more C→T substitutions than A→G substitutions. Because $a_{C \to T}$ is the ratio of the number of C→T substitutions over that of A→G substitutions, it is named a rate ratio variable to reflect this fact. For this reason, the 12 $a_{i \to j}$ values in Table 1 will yield 11 rate ratio parameters because one of the 12 will be set to 1. The number of the rate ratio parameters is one fewer than that of the rate parameters.

It should be clear to you now that a substitution model is characterized by two categories of parameters: the frequency parameters and the rate ratio parameters. The full model of nucleotide substitutions has three free frequency parameters and 11 free rate ratio parameters.

All nucleotide-based models differ from each other only in their assumptions concerning these two categories of parameters. The Jukes and Cantor's one-parameter model (Jukes and Cantor 1969) and Kimura's two-parameter model (Kimura 1980) assume equal frequency parameters, whereas all the other published models have three frequency parameters. Jukes and Cantor's model has only one rate parameter, and consequently no rate ratio parameter. Kimura's two-parameter model, the F84 model (Felsenstein 1993) and the HKY85 model (Hasegawa et al. 1985) have two rate parameters, one for transition and one for transversion. These models consequently have one rate ratio parameter, i.e., the transition/transversion ratio or κ, to estimate. The TN93 model (Tamura and Nei 1993) has three rate parameters, one for the C↔T transition, one for the A↔G transition, and one for all transversions. The ratio of transversion is typically set to one, so that the TN93 model will have two rate ratio parameters designated as κ_1 and κ_2. The REV (Yang 1994) or GTR (general time-reversible) model has six rate parameters, two for the two kinds of transitions and four for the four kinds of transversions. If the time-reversibility assumption cannot be met, e.g., when C→T transitions occurs much more frequently than T→C transitions, then we have 11 (=12 - 1) rate ratio parameters.

Given this plethora of substitution models, which one is appropriate for your sequences? A statistically rigorous approach is to do a likelihood ratio test for comparing two candidate models. For example, if you suspect that C↔T transitions occur much more frequently than the A↔G transitions, or vise versa, and therefore wish to know if the TN93 model provides significantly better fit than a simpler model, say, F84, that ignores the rate heterogeneity between C↔T and A↔G transitions, then a likelihood ratio test would help you make the decision. The likelihood ratio test and its applications will be covered in later chapters.

What you will learn in this chapter is to empirically document a substitution pattern to help you understand the frequency and the rate ratio parameters. Upon finishing this paper, you should be able to quickly identify which substitution model is appropriate for your phylogenetic analysis. Note that a substitution model is an abstraction of the true substitution pattern. A substitution pattern can be observed, although in a distorted way because of multiple hits at the same site. One of the objectives of the chapter is to help you gain an intuitive appreciation of the frequency parameters and rate ratio parameters in most substitution models. You will also understand that people did not present more and more complicated models for no good reason.

2. USE DAMBE TO DOCUMENT EMPIRICAL SUBSTITUTION PATTERNS

DAMBE provides two kinds of output for pair-wise nucleotide differences between sequences, one **simple** and one **detailed**. The **simple** output includes the number of different nucleotide sites between the two sequences, the number of nucleotide differences per nucleotide site, and the Jukes and Cantor's (1969) distance. This output is implemented for teaching purposes only, so that students can appreciate the effect of correcting for multiple hits at the same sites. The **detailed** output breaks the nucleotide differences into two kinds of transitions and four kinds of transversions.

2.1 Simple output

Start DAMBE, and open a sequence file (if you have not done so). From the menu, click **Seq. Analysis|Nucleotide Difference|Simple Output**. The standard file **Open/Save** dialog box appears. Type in the file name for saving the result, or simply use the default. Then click the **Save** button. The file is saved in text format, and a partial sample output (for only two pair-wise comparisons) is shown below:

```
Output from pairwise comparisons for sequences in file:
C:\MS\DAMBE\Diverse.fas
N.Diff - the number of nucleotide differences between the two
sequences.
Raw D - N.Diff divided by sequence length.
JC - Jukes and Cantor's distance

                                  N.Diff     Raw D         JC
XELEF1A1 vs. XELEF1ALA                 3    0.0022     0.0022
XELEF1A1 vs. XLEF1A                   47    0.0344     0.0352
XELEF1A1 vs. MMEF1A                  215    0.1572     0.1764
XELEF1A1 vs. MUSMS1X                 292    0.2135     0.2512
XELEF1A1 vs. XLEF1AB                 280    0.2047     0.2390
XELEF1A1 vs. ASEF1A                  327    0.2390     0.2878
XELEF1A1 vs. MRTEF2                  349    0.2551     0.3118
```

The output is self-explanatory. The left of the output shows the sequence pairs, with the number of possible sequence pairs being N*(N-1)/2, where N is the number of sequences. For example, if you have 8 sequences, then the number of possible sequence pairs is 8*(8-1)/2 = 28. One can see from the output that when pair-wise nucleotide differences are small, Jukes and Cantor's distance is similar to Raw D, which is simply the percentage difference. However, when the differences increase, then the saturation effect begins to show, and the JC distance, which corrects for multiple hits, becomes larger than Raw D.

2.2 Detailed Output

2.2.1 The Observed Substitution Pattern

Start DAMBE, and open a sequence file (if you have not done so). From the menu, click **Analysis|Nucleotide Difference|Detailed Output**. A dialog box appears with two lists (Fig. 1). The one on the left has the sequence pairs that are available for selection. Each number represents one sequence in the file, with number 1 representing the first sequence in the file, number 2 the second, number 3 the third, and so on. The one on the right lists sequence pairs selected for computing pair-wise nucleotide differences between the two sequences. At this moment, the list on the right is empty.

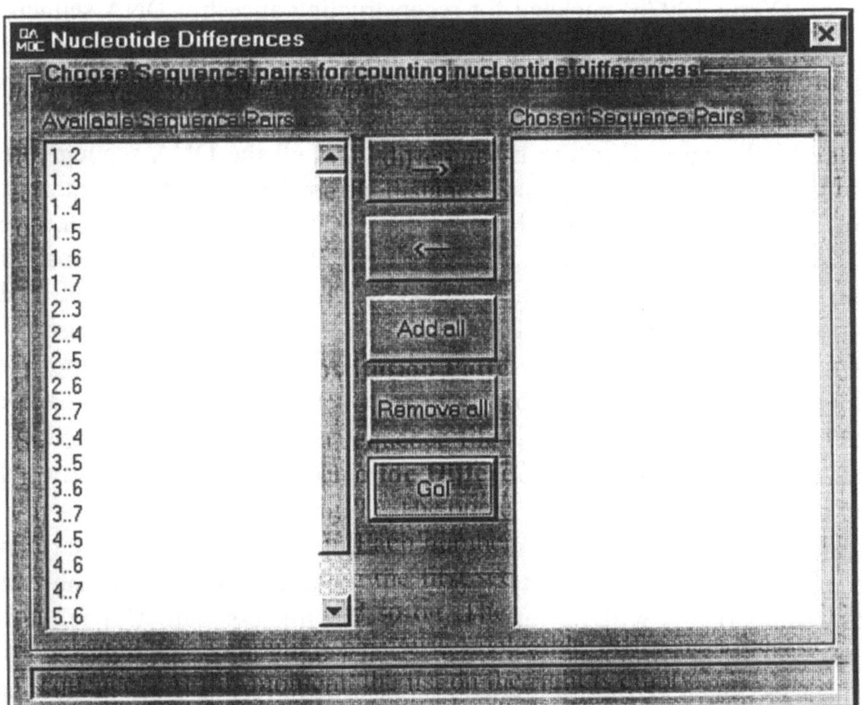

Figure 1. Dialog box for selecting sequences for pair-wise comparisons, when the input file is not in RST format, i.e., there is no phylogenetic information in the input file.

The sequence pairs available for selection on the left list depend on what input file format you use. If your input format is NOT the RST format (the output file generated by DAMBE when you choose to reconstruct ancestral sequences by using the maximum likelihood method), then the number of possible sequence pairs is simply N*(N-1)/2. The substitution patterns derived from such pair-wise comparisons may introduce biases because the comparisons are not independent (Felsenstein 1992; Nee et al. 1996; Xia et al. 1996). For example, if there is one species that has recently experienced a large number of A→G transitions and few other substitutions, then all pair-wise comparisons between this species and the other species will each contribute one data point with a large A→G transition bias.

One way to avoid such a problem of non-independence is to reconstruct ancestral states of DNA sequences and estimate the number of substitutions between neighboring nodes on the phylogenetic tree (Gojobori et al. 1982; Tamura and Nei 1993; Xia 1998b; Xia et al. 1996; Xia and Li 1998). This approach requires a phylogenetic tree for reconstructing ancestral states of internal nodes. DAMBE uses the codes in the BASEML program in the PAML (Yang 1997b) package for reconstructing ancestral DNA sequences. The output file includes a tree topology together with DNA sequences for all terminal taxa and internal nodes (the latter being reconstructed from sequences for the terminal nodes).

DAMBE can read in the tree topology and all the DNA sequences (for terminal taxa and the internal nodes) and make pair-wise comparisons between neighboring nodes along the tree. For a rooted tree with 8 species (DNA sequences), there are a total of 14 (=8*2 – 2) pair-wise comparisons, with 8 comparisons between terminal nodes and internal nodes, and 6 comparisons between internal nodes. These 14 pair-wise comparisons are independent of each other, in contrast to all possible pair-wise comparisons which amounts to 28. This topology-based pair-wise comparison has been used in studying transition bias (Xia et al. 1996), rate heterogeneity over sites (Xia 1998b), and the relative importance of amino acid properties (Xia and Li 1998).

The output from DAMBE incorporating the tree topology will be referred hereafter as the tree-based output, and the output involving all possible pair-wise comparisons will be referred to as the treeless output. The treeless output always results from pair-wise comparisons between terminal sequences, whereas the tree-based output always results from comparisons between terminal sequences and their immediate ancestors, or between ancestral sequences.

Regardless of which input file format you use, the selection of sequence pairs is similar to what you have done with getting nucleotide frequencies. Once you have finished your selection, click the **Done** button. After a few seconds, the standard **file/open** dialog box appears. Type in the file name for saving the result, or simply use the default. Then click the **Save** button. The file is saved in text format, and a partial sample output for a set of elongation factor 1-α sequences (for only one pair-wise comparison between two chelicerate species) is shown below:

```
Chelicer1 vs. Chelicer2
==========================================================================
        ------Identical---------  -Transition-   -----Transversion-------
        AA     CC     GG     UU     AG     CU     AC     AU     CG     GU
Obs    259    190    227    205     44     74     26     35      8     12
Exp    291    178    201    211     37     30     35     38     29     31
==========================================================================

    A   G   C   T
A   0
G  44   0
C  26   8   0
T  35  12  74   0

     A    G    C    T    SubTotal Prop
A  259   10    7   11       287   0.266
G   34  227    5    8       274   0.254
C   19    3  190   54       266   0.246
T   24    4   20  205       253   0.234

S  336  244  222  278     1,080
P 0.31 0.23 0.21 0.26

Statistic                       Estimate      StdDev
Juke and Cantor's Distance        0.2115      0.0156
Kimura's two-parameter D          0.2144      0.0161
Tajima and Nei's (1984) D         0.2178      0.0167
Lake's Paralinear Distance        0.2141
==========================================================================
```

The sample output is produced from DNA sequences of the EF-1α gene from chelicerate species. You can see that the output is treeless because the comparisons are between terminal taxa. The result for each pair-wise comparison is presented in three parts. The first part lists the observed number of identical pairs, pairs differing by a transition and pairs differing by a transversion. The expected numbers, based on nucleotide frequencies only, is shown below the observed numbers. Nucleotides of high frequencies are expected to be involved in substitutions more frequently than rare nucleotides. In the extreme case when a nucleotide is not found in the sequences, then obviously it will not be observed in a substitution. In our data, A and T have the highest frequencies, and A↔T substitutions are expected to occur most frequently, everything else being equal.

Notice that the two types of transitions occur more frequently than expected, and the four types of transversions all occur less frequently than expected. This empirical evidence favours substitution models incorporating the s/v ratio (e.g., F84 and HKY85 models) against simpler models such as the JC69 and TN84 models.

We further notice that A↔G transitions are expected to occur more frequently than C↔T transitions because the frequencies of A and G are greater than those of C and T. However, the observed pattern is the opposite, with C↔T (i.e., C↔U) transitions far outnumbers the A↔G transitions. This empirical evidence favours the TN93 model against the simpler F84 and HKY85 models.

There is also heterogeneity among the transversions. For example, transversions involving G are rare relative to transversions involving A. This suggests that even the TN93 model may not be realistic enough and we need to have models that do not assume the same rate for all four kinds of transversions. Many researchers choose particular substitution models just because these models happen to be available in the computer program they use. This is poor practice. One should always make a judicious choice of substitution models based on empirical evidence. A number of researchers believed that the maximum likelihood method would recover the correct topology even if the underlying model is false. Such a belief is based on myth rather than reason.

The second part of the output is a summary of substitution patterns in two matrices, the lower-triangle matrix and a square matrix. These matrices allow one to quickly identify which substitution is appropriate for the sequences. For example, if nucleotide frequencies are roughly equal, but the numbers on either side of the diagonal in the square matrix are not at al symmetrical, i.e., when the matrix differ greatly from its transverse, then the assumption of time reversibility may not hold.

A substitution pattern is said to be time-reversible if the probability of forward mutation equals the backward mutation, i.e., the probability of observing an A→G transition is the same as observing a G→A transition. Many substitution models, from the simple Jukes and Cantor's (1969) one-parameter model to the complicated REV model (Yang 1994), assume time-reversibility.

The number of observed nucleotide substitutions is also broken down into the three codon positions for protein-coding genes, but the output is not reproduced here. The reader is expected to read the output and make his own discoveries. For example, it will be clear that most substitutions occur at the third codon position, fewer at the first position, and fewest at the second position.

Because the EF1-α gene is a protein-coding gene, the most appropriate substitution model should be codon-based (Goldman and Yang 1994; Muse and Gaut 1994) rather than nucleotide-based. However, the current implementation of codon-based model is not quite practical.

The third part of the output is a plot of the number of transitions and transversions versus Kimura's two-parameter distance (K_{K80}). If we take K_{K80} as a measure of divergence, then both the transitional and transversional substitutions should increase with K_{K80}, with the former increasing faster than the latter. However, with the increase of divergence time, multiple substitutions at the same site would occur, and transversions will gradually outnumber transitions. This is a visual way of detecting substitution saturation.

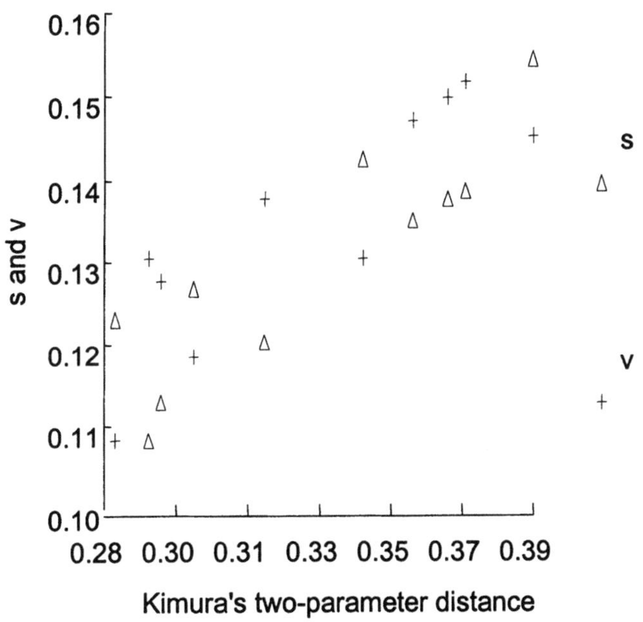

Figure 2. Plotting estimated number of transitions and transversions versus Kimura's (1980) two-parameter distances from pair-wise comparisons.

Chapter 13

Preamble to the Pattern of Codon Substitution

The default pattern when there is no purifying selection

1. INTRODUCTION

In a previous chapter dealing with codon frequencies, we have developed basic appreciation of factors affecting codon frequencies. In this chapter and the one that follows, we study the rate of codon substitution and factors affecting the rate of codon substitutions. We know that substitutions are affected by both mutation and selection, and the observed codon usage pattern and codon substitution pattern are the results of mutation-selection balance. Before we use DAMBE to study observed patterns of codon substitutions, it is helpful for us to develop the expected codon substitution pattern when there is no purifying selection.

You will find that such an attempt will help us to resolve some old controversies between Kimura (1983, p. 159) and Gillespie (1991, p. 43). Kimura, being a neutralist, argued that the most frequent nonsynonymous substitutions were those involving similar amino acids and the substitution rate would decrease as involved amino acids were more different (fig. 7.1 in Kimura 1983). This is of course what one would expect from the neutral theory of molecular evolution, in which positive selection plays a negligible role in molecular evolution and purifying (negative) selection eliminates those mutations with major effects. Gillespie, on the other hand, argued that the most frequent nonsynonymous substitutions are not between the chemically most similar amino acids, but instead are between a group of amino acids with a Miyata's distance (Miyata et al. 1979) near one (fig. 1.12 in Gillespie 1991). In short, Kimura found the substitution pattern consistent with the effect of purifying selection whereas Gillespie found the evidence consistent with positive selection.

Both Kimura and Gillespie based their conclusions on a plot of observed rate of amino acid substitutions versus amino acid dissimilarity. What would the plot be like if there is no selection at all? There are 196 possible nonsynonymous codon substitutions involving a single nucleotide substitution. How many of them involve very similar or very different amino acids? Is it likely that we will obtain substitution patterns similar to those documented by Kimura or Gillespie even when there is no selection at all?

2. DEFAULT SUBSTITUTION PATTERNS WITH NO SELECTION

Start DAMBE and open a sequence file, e.g., **invert.rst** that comes with DAMBE. This file contains protein-coding nucleotide sequences of the elongation factor-1α gene (EF-1α) from seven invertebrate species as well as reconstructed ancestral sequences for internal nodes. The file also contains a topology used for the reconstruction of ancestral sequences. You will learn how to use DAMBE to reconstruct ancestral sequences in a later chapter on molecular phylogenetics. The resulting file from such reconstruction has a **.rst** file type by default. Such files are referred to as RST format in this book.

Click **Seq. Analysis|Expected (Freq. unadjusted)**. This requests expected codon substitution pattern with respect to amino acid dissimilarity measured by Grantham's distances. DAMBE will ask you to name a file for saving the output. Enter a name in the dialog box and click the **Save** button. You will also be asked to input a transition/transversion ratio. Just use the default value of 1. The output, which actually has nothing to do with your input sequences, is in two parts. Although only Part II is relevant to our discussion, information contained in Part I might be interesting to you, too. So let us have a quick look at it.

```
Part I:
s - transition; v - transversion; G - Grantham's distance.

                 ---Codon Position----
s/v                1          2          3      Subtotal      Prop
==================================================================
 s                27         30          1           58      0.296
Mean G         67.37      88.47      10.00        77.29
Var G        1889.70    2104.67                 2124.56
```

```
v               56         58         24        138      0.704
Mean G       69.41     102.62      73.17      84.02
Var G      2390.61    1399.08    3031.45    2305.12
============================================================
Sum             83         88         25        196
Prop         0.423      0.449      0.128        1.0
Mean G       68.75      97.80      70.64
Var G      2203.56    1663.73    3064.74
============================================================
Mean G for all:      82.031
Var G for all:     2249.999
```

The expected pattern of codon substitution tells us that 42.3% of the SSNCS (i.e., single-step nonsynonymous codon substitutions) should fall at the first codon position, 44.9% at the second, and only 12.8% at the third. Similarly, we expect 29.6% of SSNCS to be transitions and 70.4% to be transversions. You will learn in the next chapter that the observed SSNCS differ quite much from this expectation.

The second part of the output, reproduced below, shows the distribution of the 196 possible SSNCS over Grantham's distances. For example, there are four SSNCS that have Grantham's distance equal to 5. The third column is simply the frequency value divided by the total of 196. Note that the total will be different if you use a different genetic code. For example, there will be only 190 possible SSNCS for mammalian mitochondrial code.

```
Part II:
=====================================
Grantham's D   Frequency  Proportion
          5           4      0.0204
         10           3      0.0153
         14           2      0.0102
         21           2      0.0102
         22           9      0.0459
         23           2      0.0102
         26           2      0.0102
         27           4      0.0204
         29           4      0.0204
         30           3      0.0153
         32           6      0.0306
         38           4      0.0204
         41           4      0.0204
         46           2      0.0102
         50           2      0.0102
         53           2      0.0102
         54           2      0.0102
```

55	2	0.0102
57	2	0.0102
58	10	0.0510
60	4	0.0204
61	5	0.0255
64	4	0.0204
65	2	0.0102
68	2	0.0102
71	2	0.0102
73	4	0.0204
76	2	0.0102
78	2	0.0102
81	3	0.0153
83	2	0.0102
87	2	0.0102
89	3	0.0153
91	2	0.0102
92	1	0.0051
94	9	0.0459
96	2	0.0102
98	5	0.0255
99	6	0.0306
101	2	0.0102
102	5	0.0255
103	4	0.0204
109	10	0.0510
112	4	0.0204
125	6	0.0306
126	2	0.0102
138	2	0.0102
142	4	0.0204
143	2	0.0102
144	2	0.0102
149	2	0.0102
152	2	0.0102
155	2	0.0102
158	2	0.0102
160	2	0.0102
177	1	0.0051
180	2	0.0102
184	1	0.0051
194	2	0.0102
204	2	0.0102
214	2	0.0102

DAMBE will ask you if you wish to obtain a plot of the **Proportion** versus Grantham's distance. Click **Yes** to generate the plot (fig. 1). Few of the 196 possible SSNCS have a large Grantham's distance. In other words, the genetic code has constrained the codon substitution in such a way that a single nucleotide mutation will most likely lead to either a nonsynonymous substitution (in which case Grantham's distance = 0) or to an amino acid similar to the original. We should expect to see few nonsynonymous codon substitutions of large effects even if all 196 SSNCS occur equally frequently, with no involvement of any selection. It is unnecessary for Kimura to invoke purifying selection to explain the rarity of nonsynonymous substitutions of large effects.

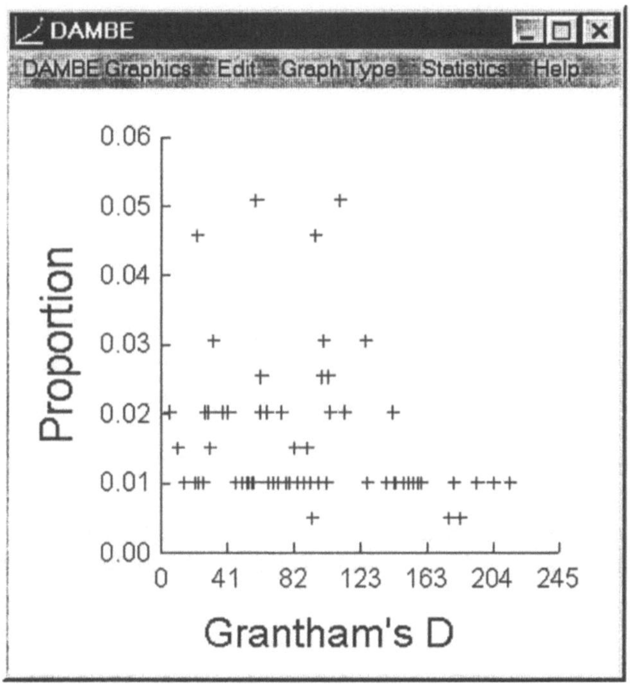

Figure 1. The distribution of Grantham's distances for the 196 possible single-step nonsynonymous codon substitutions based on the universal genetic code.

We also notice that most of these 196 SSNCS have a Grantham's D near 82, which is equivalent to Miyata's distance of 1. In other words, the plot produced by Gillespie does not really need to invoke positive selection to explain.

The relationship between **Proportion** and **Grantham's D** is not adjusted for codon frequencies. However, the adjustment makes very little difference.

You can verify that by clicking **Seq. Analysis|Expected P123 (Freq. Adjusted)**. A sample plot with the expected substitutions adjusted for the codon frequencies in the input sequences in the **invert.rst** file is shown in fig. 2. Note that it conforms even better to Gillespie's selectionist scenario than fig. 1. Both Kimura and Gillespie did adjust their data according to amino acid frequencies.

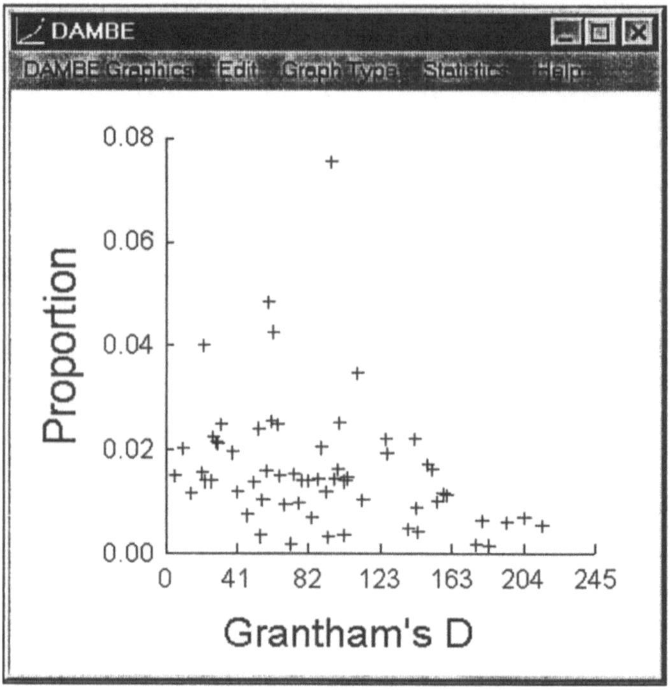

Figure 2. The distribution of Grantham's distances for all possible single-step nonsynonymous codon substitutions based on the universal genetic code, adjusted for codon frequencies.

Chapter 14

Factors Affecting Codon Substitutions

1. INTRODUCTION

In previous chapters we developed an empirical appreciation of biased codon usage and factors affecting codon usage as well as codon substitution. Also covered in previous chapters are some controversies concerning nonsynonymous substitution and how an understanding of possible codon substitutions can help us to have a better appreciation of evolutionary dynamics of codon sequences. In this chapter we study further the rate of codon substitution and factors affecting the rate of codon substitutions. DAMBE features a number of functions that can help us to reveal the dynamic nature of codons during evolution.

1.1 The Rate of Codon Substitutions and its Determinants

The likelihood of a codon substitution depends mainly on the difference between the original codon and the replacement codon. The difference between two codons depends on two factors. One is the number of codon positions at which they differ. For example, everything else being equal, substitutions between codons differing by one codon position is more likely to occur than between codons differing by three codon positions. Thus, if we designated N_{ij} (= 0, 1, 2, 3) as the number of codon positions differing between codons i and j, then the substitution rate between codons i and j (R_{ij}) should be a decreasing function of N_{ij}.

Another difference between two codons is reflected in the amino acids they code for. For example, a codon coding for amino acid phenylalanine (e.g., UUU) is more likely to be replaced by a tyrosine codon (e.g., UAU) than by a serine codon (e.g., UCU) although in both cases the replacement involves a single nucleotide change. The reason behind this is that phenylalanine is similar to tyrosine in physico-chemical properties, but very different from serine codons. It has been experimentally shown that the replacement of tyrosine by phenylalanine (both having an aromatic ring on the side chain) at position 2 of oxytocin has little effect on the oxytocic activity of the peptide (Boissonnas and Guttmann 1960; Jaquenoud and Boissonnas 1959), but the replacement by serine (lacking the aromatic ring) reduces the oxytocic activity to an undetectable level (Guttmann and Boissonnas 1960). Thus, if we designated D_{ij} as the amino acid dissimilarity between the two amino acids coded for by codons i and j, then the substitution rate between the two codons should be a decreasing function of D_{ij} (Clarke 1970; Epstein 1967; Grantham 1974; Kimura 1983, p. 152; Miyata et al. 1979; Sneath 1966; Xia and Li 1998; Zuckerkandl and Pauling 1965).

There are two common measures of amino acid dissimilarity, one being Grantham's (1974) distance and the other being Miyata's distance (Miyata et al. 1979). The former is based on three physico-chemical properties of amino acids: volume, polarity and chemical composition of the side chain; the latter is based on the first two of the three properties. Both have some undesirable aspects (Xia and Li 1998) and better measures of amino acid dissimilarity needs to be developed. Both distances are implemented in DAMBE.

In summary, we learned that the rate of substitution between codons i and j (R_{ij}) is affected by both N_{ij} and D_{ij}, with R_{ij} being a decreasing function of both N_{ij} and D_{ij}.

1.2 Models of Codon Substitution

Although it is not difficult to see that R_{ij} is affected by both N_{ij} and D_{ij}, it is difficult to arrive at a good functional relationship between R_{ij} and the two independent variables. There are two complications. The first is that N_{ij} and D_{ij} are correlated, which you can immediately verify as follows. Start DAMBE and open a protein-coding nucleotide sequences. Click **Seq. Analysis|AA distance vs. nuc. sites different**. The following result is generated:

```
Relationship between amino acid dissimilarity and
the number of codon sites different.

Dg - Grantham's distance
Dm - Miyata's distance
```

```
Part I: Output unadjusted for codon frequencies:
=====================================================
                         Codon Site
                    1           2           3
N                 196         770         777
Mean Dg        82.031      90.674     104.560
Var  Dg      2249.999    1913.021    1403.948

N                 196         770         777
Mean Dm         1.792       2.034       2.409
Var  Dm         1.156       1.100       0.733
=====================================================
```

For the universal genetic code, there are 196 possible codon mutations involving a single codon sites, 770 involving two, and 777 involving three codon sites. The mean Grantham's distance is 82.031 for codon replacements with $N_{ij} = 1$, 90.674 with $N_{ij} = 2$, and 104.560 with $N_{ij} = 3$. Thus, the more sites by which the two codon differ from each other, the greater the amino acid dissimilarity. The difference in mean Grantham's distance or mean Miyata's distance is highly significant ($P = 0.0000$) among the three groups.

The second complication of incorporating N_{ij} into the model is that its observed effect decreases with divergence time. When you compare two sequences, the number of codon substitutions will be less dependent on N_{ij} when the two sequences have diverged for a long time. For such sequences, the number of codon substitutions depends almost entirely on D_{ij}.

Two models of codon substitution have been proposed, one incorporating the effect of D_{ij} (Goldman and Yang 1994) and the other (Muse and Gaut 1994) does not. The former, referred to as the GY94 model, has been implemented in the PAML package (Yang 1997b) for phylogenetic reconstruction and model testing, and the latter implemented in the program CODRATES for relative rate tests. The GY94 model has recently been revised (Yang et al. 1998), and the revised model will be referred to as YNH98 model.

Codon-based thinking is often not expressed in a model of codon substitution. For example, all methods that estimate the number of synonymous and nonsynonymous nucleotide substitutions (Li 1993; Li et al. 1985a; Miyata and Yasunaga 1980; Nei and Gojobori 1986; Perler et al. 1980) used codon as a unit of substitution. These distance measures, at least for the time being, makes better use of D_{ij} and N_{ij} than the two explicitly presented codon-based models. For example, two codons differing by two codon positions can be treated by these methods as two possible evolutionary pathways.

A good codon-based model, or codon-based thinking, is potentially better than a nucleotide-based model for the reason that the latter makes use of only the information in N_{ij} but ignores the information in D_{ij}. Take codons GAU (coding for Asp) and GGU (coding for Gly) for example. The third codon position (U) in both codons would be treated as equivalent in a nucleotide-based model and expected to have the same substitution rate. However, a U→A substitution in the former is nonsynonymous and occurs rarely, whereas the same U→A substitution in the latter are synonymous ($D_{ij} = 0$) and occurs relatively frequently.

You might also note that almost all nucleotide-based models assume that nucleotide substitutions occur independently among sites. In other words, whether a nucleotide substitution will occur at site i is independent of what nucleotide is found at site j. This is often not the case. In the example above involving codons GAU and GGU, whether a U→A substitution will occur depends much on whether the second position is an A or a G. The U→A substitution is rare when the second codon position is A, but relatively frequent when the second codon position is G. This violates the assumption that nucleotide substitutions occur independently among sites.

A codon-based model is also potentially better than an amino acid-based model because the latter makes use of only information in D_{ij} but ignores most of the information in N_{ij}. This disadvantage would be particularly pronounced with recently diverged lineages.

To be fair, nucleotide-based or amino acid-based models also have advantages over codon-based models. First, they have fewer parameters to estimate than the codon-based models, and consequently would be expected to have more robust estimates when sequences are short. Second, the former require less computation time than the latter. Also, because D_{ij} and N_{ij} are positively correlated, the disadvantage I mentioned above is reduced. In other words, if the information in one of the two variables, i.e., D_{ij} and N_{ij}, is incorporated in the model, then the information in the other is at least partially incorporated.

1.3 The Expected Pattern of Nonsynonymous Codon Substitutions With No Purifying Selection

DAMBE can help you to describe the observed pattern of codon substitutions. However, the observed pattern would make little sense if we do not have some expectations. Let us use an example from our daily life. When we see a baby lamb, we expect it to have parents, otherwise we would find ourselves intrigued and wish to come up with some explanations or theories, e.g, the lamb might have been a clone. The theories will then guide us to do some experiments, which might prove that the lamb is indeed a clone. If we do not have any expectation, then we will be mentally passive and will not be involved in active learning.

Let us now develop some basic expectations of nonsynonymous codon substitutions when there is no purifying selection. For illustration, we will use mammalian mtDNA to begin with. Of the 60 mitochondrial codons, there are 190 possible nonsynonymous codon pairs in which one codon can mutate into the other through a single nucleotide substitution, e.g., ACU-GCU (Reciprocal codon pairs, e.g., ACU-GCU and GCU-ACU, were treated as the same type of nonsynonymous codon substitutions, otherwise there would have been 380 possible nonsynonymous codon pairs differing at one codon position.) These 190 nonsynonymous codon pairs are grouped into five categories according to whether the nonsynonymous substitution occurs at the first, second or third codon position, and whether it is a transition or transversion. The result (Table 1) shows that, when we compare two DNA sequences and count nonsynonymous codon pairs that differ at one codon position, we should expect, assuming equal codon usage and equal probability of nonsynonymous substitutions, 43.2% (=82/190) of the nonsynonymous codon pairs to differ at the first codon position, 44.2% at the second codon position, and only 12.6% at the third codon position. Similarly, we should expect 28.4% of nonsynonymous codon pairs to differ by a transition, and 71.6% to differ by a transversion (Table 1).

The expected pattern (Table 1) can be directly generated for any of the 12 genetic codes implemented in DAMBE. Click **Sequence|Change Sequence Type** to choose appropriate genetic code, and then click **Seq. Analysis|Expected P123 (Freq. unadjusted)**. You will see Table 1 right in the display window.

Table 1. Expected pattern of nonsynonymous codon substitutions for mammalian mitochondrial DNA, with respect to codon position and transition/transversion. D_G - Grantham's distance.

	Codon position			Subtotal	Proportion
	1	2	3		
s	26	28	0	54	0.284
Mean D_G	63.92	92.64	Null	78.81	
Var D_G	1758.95	1955.5	Null	2035.7	
V	56	56	24	136	0.716
Mean D_G	71.21	104.46	73.67	85.34	
Var D_G	2568.68	1541.13	4879.54	2764.4	
Sum	82	84	24	190	
Prop.	0.432	0.442	0.126	1	
Mean D_G	68.9	100.52	73.67		
Var D_G	2298.71	1688.78	4879.54		

The expected pattern of nonsynonymous codon substitution shown in Table 1 assumes equal usage of the sense codons. In reality, some codons are used much more frequently than others. For example, idiosyncratic amino acids (e.g, Cys, Trp) and their codons are used less frequently than "typical" amino acids such as Leu and Thr and their codons (Xia and Li 1998). Thus it seems necessary to adjust for codon usage. Such adjustment, however, only results in minor differences (Xia 1998b). You can verify this by open a file containing protein-coding nucleotide sequences and obtain the equivalent of Table 1 adjusted for codon frequencies by clicking **Seq. Analysis|Expected P123 (Freq. Adjusted).**

If there is strong purifying selection, then those nonsynonymous codon substitutions resulting in replacement of very different amino acids will be selected against, and only those nonsynonymous codon substitutions resulting in replacement of very similar amino acids will have a chance to be fixed and observed.

2. CODON COMPARISON WITH DAMBE

2.1 Tracing evolutionary history

We often see evolutionary information, such as the number of nonsynonymous substitutions superimposed on a phylogenetic tree (fig. 1). The values on each branch (fig. 1) is obtained by first constructing ancestral sequences and then perform pair-wise comparisons between neighboring nodes along the tree. Why such branch-specific information is important? One particular way of using the branch-specific information is to correlate the information, e.g., the number of synonymous or nonsynonymous changes with phenotypic characters. For example, a neutralist would argue that most substitutions at the nucleotide or amino acid level are nearly neutral and the number of changes of synonymous or nonsynonymous substitutions would have little to do with changes in phenotypic characters. In contrast, a selectionist would argue that a substantial fraction of substitutions, especially nonsynonymous substitutions, may have phenotypic consequences. For example, a large number of nonsynonymous substitutions along one particular branch may result in a corresponding change in morphological, physiological or behavioural changes. The comparative method (Felsenstein 1985c; Felsenstein 1988a; Harvey and Pagel 1991) allows one to partition the observed variation of any particular phenotypic character along the branches so that correlation between any particular phenotypic trait such as developmental symmetry, and any particular genetic character such as nonsynonymous substitutions can be calculated and tested for significance.

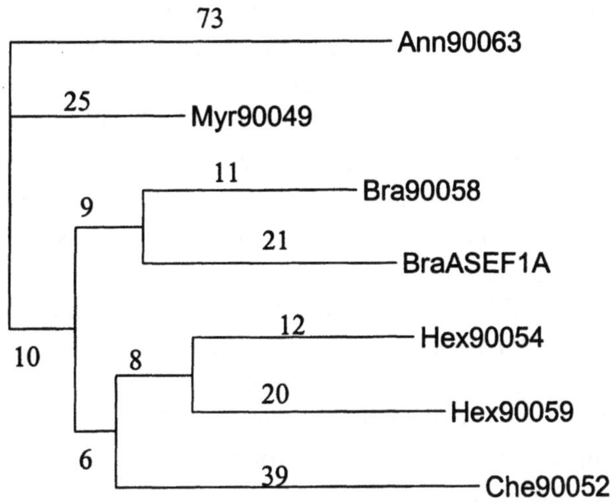

Figure 1. Number of nonsynonymous substitutions superimposed on a phylogenetic tree of seven invertebrate species

Sometimes is not sufficient to put just a number of each branch of a tree. One may need to know exactly which kind of nonsynonymous substitutions have occurred at which position. DAMBE provides a convenient way for you to generate different kinds of branch-specific information.

Start DAMBE, and open a protein-coding sequence file, e.g., the **invert.rst** file that comes with DAMBE. This file contains protein-coding nucleotide sequences of the elongation factor-1α gene (EF-1α) from seven invertebrate species as well as reconstructed ancestral sequences for internal nodes. The file also contains a topology used for the reconstruction of ancestral sequences. You will learn how to use DAMBE to reconstruct ancestral sequences in a later chapter on molecular phylogenetics. The resulting file from such reconstruction has a **.rst** file type by default. Such files are referred to as RST format in this book.

Click the **Seq. Analysis|Codon Difference|Nonsynonymous codon substitution with AA dissimilarity**. A dialog box appears for you to choose which sequence pairs you wish to perform pair-wise comparisons. The listbox on the left has the sequence pairs that are available for selection. Each number represents one sequence in the file, with number 1 representing the first sequence in the file, number 2 the second, number 3 the third, and so on. The listbox on the right lists sequence pairs selected for computing pair-wise nucleotide differences between the two sequences, and is empty at the very beginning.

The sequence pairs available for selection on the left listbox depend on what input file format you use. If your input format is NOT the RST format, then the number of possible sequence pairs is simply N*(N-1)/2, where N is the number of sequences. The substitution patterns derived from such pair-wise comparisons may introduce biases because the comparisons are not independent (Felsenstein 1992; Nee et al. 1996; Xia et al. 1996). For example, if there is one species that has recently experienced a large number of A→G transitions and few other substitutions, then all pair-wise comparisons between this species and the other species will each contribute one data point with a large A→G transition bias.

One way to avoid such a problem of non-independence is to reconstruct ancestral states of DNA sequences and estimate the number of substitutions between neighboring nodes on the phylogenetic tree (Gojobori et al. 1982; Tamura and Nei 1993; Xia 1998b; Xia et al. 1996; Xia and Li 1998). This approach requires a phylogenetic tree for reconstructing ancestral states of internal nodes. DAMBE used the codes in the BASEML program in the PAML (Yang 1997b) package for reconstructing ancestral DNA sequences. The output file includes a tree topology together with DNA sequences for all terminal taxa and internal nodes (the latter being reconstructed from sequences for the terminal nodes).

DAMBE can read in the tree topology and all the DNA sequences (for terminal taxa and the internal nodes) and make pair-wise comparisons between neighboring nodes along the tree. For a rooted tree with 8 species (DNA sequences), there are a total of 14 (=8*2 – 2) pair-wise comparisons, with 8 comparisons between terminal nodes and internal nodes, and 6 comparisons between internal nodes. These 14 pair-wise comparisons are independent of each other, in contrast to all possible pair-wise comparisons which amounts to 28. This topology-based pair-wise comparison has been used in studying transition bias in mammalian mitochondrial genes (Xia et al. 1996), rate heterogeneity over sites (Xia 1998b), and the relative importance of amino acid properties on protein evolution (Xia and Li 1998). The output from DAMBE incorporating the tree topology will be referred hereafter as the tree-based output, and the output involving all possible pair-wise comparisons will be referred to as the treeless output. The treeless output always results from pair-wise comparisons between terminal sequences, whereas tree-based output always results from comparisons between terminal sequences and their immediate ancestors, or between ancestral sequences.

Regardless of which input file format you use, the selection of sequence pairs is similar to what you have done with getting nucleotide frequencies.

Once you have finished your selection, click the **Done** button. DAMBE will compare each sequence pairs codon by codon and display information on nonsynonymous codon pairs. After a few seconds, the standard **file/open** dialog box appears. Type in the file name for saving the result, or simply use the default. Then click the Save button. The file is saved in text format, and a partial sample output for a set of elongation factor 1-α sequences (for only one pair-wise comparison between two chelicerate species) is shown below:

```
N      Cod1    Cod2    AA1  AA2    DG      DM
============================================

node#8 vs. node#9

 37    AGG     AGU     R    S   109.00    2.73
 106   ACC     GCC     T    A    58.00    0.90
 150   UCA     CCA     S    P    73.00    0.55
 241   AAG     GAG     K    E    53.00    1.05
 244   GUU     CUU     V    L    32.00    0.91
 357   GAA     GAC     E    D    61.00    1.46
--------------------------------------------
Mean                             64.33    1.27
Num NS: 6

.....

node#11 vs. Bra90058

 26    UAC     UUC     Y    F    22.00    0.48
 106   GCC     AAC     A    N   110.00    1.77
 117   ACU     UCU     T    S    58.00    0.89
 147   GCC     ACC     A    T    58.00    0.90
 148   AAG     AAC     K    N    94.00    1.83
 149   AUG     UUG     M    L    14.00    0.41
 156   AAC     GCC     N    A   110.00    1.77
 171   GAA     GAC     E    D    61.00    1.46
 213   AUG     AUC     M    I    10.00    0.29
 272   AAC     AGC     N    S    46.00    1.31
 282   UCU     UAC     S    Y   143.00    3.32
--------------------------------------------
Mean                             66.00    1.31
Num NS: 11
```

Pair-wise comparisons along the tree are either between internal nodes, or between an internal node and a terminal node. This information is shown at the beginning of each pair-wise comparisons. The first column shows the sequential numbering of codons along the DNA sequences (after deleting unresolved codons). The second and third columns show which codon pairs are involved in the substitution, and the fourth and fifth columns show the corresponding amino acids.

The second and the last columns are, respectively, the Grantham's and Miyata's distances between the two amino acids. Grantham's distance is scaled to have a mean of 100, so that, if amino acid substitutions are random, then the mean Grantham's distance should be roughly 100. The mean of the Grantham's distances in the output is significantly smaller than 100 ($P < 0.001$), by either the parametric t-test or the nonparametric Mann-Whitney's test or sign test.

The number of nonsynonymous substitutions is also presented in the output. This number is what we can superimpose onto a phylogenetic tree (fig. 1). Don't quit DAMBE yet.

2.2 Summary of codon substitution pattern

While the **invert.rst** file is still in DAMBE's buffer, click **Seq. Analysis|Codon Difference|Summary output**. DAMBE will carry out pair-wise, codon by codon comparisons and produce a summary of the codon substitution pattern for each pair of sequences compared. An overall summary is also generated at the end of the output and is partially reproduced below:

Codon Pair	DiffID	Freq.	Prop.	Mean G	Mean M
Identical		3046			
Synonymous	100	26	0.036		
	010	0	0.000		
	001	650	0.912		
	110	8	0.011		
	101	27	0.038		
	011	0	0.000		
	111	2	0.003		
Sum		713	1.0		
Nonsynonymous	100	70	0.299	49.21	0.84
	010	28	0.120	70.50	1.45
	001	33	0.141	58.27	1.36
	110	27	0.115	97.22	1.90
	101	33	0.141	40.64	0.68
	011	27	0.115	73.07	1.76
	111	16	0.068	108.81	2.11

```
Sum                    234  1.0
```

The output might look a bit confusing at first, but you will soon find it quite simple. There are three types of codon pairs from the pair-wise comparisons: identical (N = 3046), synonymous (N = 713) and nonsynonymous (N = 234). The details of the 234 nonsynonymous substitutions can be obtained by using the method introduced in the previous section. The three digits under **DiffID** refers to codon positions 1, 2, and 3, with a "1" meaning different nucleotides and a "0" meaning an identical nucleotide. If two codons differ by codon position 1 and 3, DiffID will be written as 101.

Of the 713 synonymous codon substitutions, most involve a change at the third codon position (N = 650). This is because the probability of a nucleotide substitution at the third codon position being synonymous is by far the largest compared to a substitution at the first and second codon position. The probability of a synonymous substitution is in fact zero for substitutions occurring at the second codon position.

There are two patterns that you can recognize. First, most codon substitutions are synonymous and, for nonsynonymous codon substitutions, those resulting in replacement of similar amino acids, with small Grantham's distances, occur more frequently than those resulting in replacement of very different amino acids. For example, among the codon pairs that differ by a single nucleotide, those differ at the first codon position have occurred most frequently and also have the smallest Grantham's distance (= 49.21) and Miyata's distance (= 0.84). This pattern is also true for codon pairs that differ at two codon positions, with those differing at the first and third codon positions occurring most frequently and having the smallest Grantham's and Miyata's distances. Thus, the effect of D_{ij} (the difference in physico-chemical properties between amino acids), plays a significant part in determining the rate of codon substitution. Second, most codon substitutions involve codon pairs differing at one codon position (N = 807), fewer differing at two codon positions (N = 122), and still fewer differing at all three positions (N = 18). This illustrates the effect of N_{ij} (the number of codon positions by which two codons differ), on the rate of substitution.

Of the 234 nonsynonymous substitutions, 131 involve codon pairs differing at a single codon position. We will have a closer look at these 131 "single-step" nonsynonymous codon substitutions in the next section. Please continue on.

2.3 Single-step Nonsynonymous Codon Substitutions

While the **invert.rst** file is still in DAMBE's buffer, click **Seq. Analysis|Codon difference|One-step nonsyn codon substitution**. DAMBE will perform pair-wise comparisons and pick up all those nonsynonymous substitutions involving a single nucleotide change. The summary output is in two part, with Part I reproduced below:

```
Part I:
                 ---Codon Position----
s/v                 1         2         3     Subtotal      Prop
==============================================================
 s                 28         9         1          38      0.290
Mean G          46.00     94.44     10.00       56.53
Var G          303.04   2558.53               1266.42

 v                 42        19        32          93      0.710
Mean G          51.36     59.16     59.78       55.85
Var G         1104.53   1212.81   1107.79     1119.65
==============================================================
Sum                70        28        33         131
Prop            0.534     0.214     0.252         1.0
Mean G          49.21     70.50     58.27
Var G          781.88   1848.26   1148.27
```

Remember that, of the 234 nonsynonymous codon substitutions, 131 involve codon pairs differing at one codon position, which is often termed SSNCS (for single-step nonsynonymous codon substitution). The output consists of two parts. Part I of the output breaks down these SSNCSs according to codon positions and whether it is a transition or transversion. We see that 28 SSNCSs involve a transition at the first codon position, with a mean Grantham's distance of 46, two involve a transversion at the first codon position, with a mean Grantham's distance of 51.36, and so on.

These values by themselves make little sense. However, we can compare these observed pattern of SSNCS with the expected pattern. If DAMBE is still running and your sequences are still in the buffer, click **Seq. Analysis|Expected (Freq. adjusted)**. This will generate the expected pattern of nonsynonymous codon substitutions involving a single nucleotide change, when there is no purifying selection. The differential codon usage has already been adjusted for. DAMBE will ask you to input a name for saving the output. Enter a name in the dialog box and the output will be saved in text format. The output, with its format similar to the previous sample output, is partially shown below:

s/v	---Codon Position---- 1	2	3	Subtotal
s	0.142	0.150	0.003	0.295
Mean G	56.82	84.33	10.00	70.24
Var G	1266.58	1606.67		1653.63
v	0.273	0.298	0.133	0.705
Mean G	65.71	102.55	65.18	81.20
Var G	2249.53	1314.24	1917.54	2125.22
Sum	0.415	0.448	0.137	1.0
Mean G	62.67	96.45	63.81	
Var G	1930.87	1485.79	1943.79	

```
Mean G for all:      77.955
Var G for all:     2001.603
```

The expected pattern of codon substitution tells us that 41.5% of the SSNCSs should fall at the first codon position, 44.8% at the second, and only 13.7% at the third. The observed SSNCS has 53.4%, 21.4%, and 25.2% falling at the first, second and third codon positions, respectively. In other words, there is a surplus of SSNCSs at the first and third codon positions, and a deficiency of SSNCSs at the second codon position. This deviation from the expected pattern, tested by a chi-square test, is highly significant (P = 0.000).

The pattern can be easily explained by examining the mean Grantham's distance at the three codon positions, which is 62.67, 96.45, and 63.81, respectively, for the first, second and third codon positions. Nonsynonymous substitutions at the second codon position apparently involve amino acid replacements of larger effects than those at the first or the third codon position. In other words, nonsynonymous substitutions at the second codon position are subject to stronger purifying selection, which is why we see fewer nonsynonymous substitutions at the second codon position than expected.

Part II of the output, reproduced below, deserves a closer look. DAMBE plots the observed frequency of nonsynonymous substitutions (the column headed by **Proportion** in the output, designated hereafter as P_{obs}) versus Grantham's distance (fig. 2) which measures amino acid dissimilarity. The plot shows that nonsynonymous substitutions involving very different amino acids (i.e., a large Grantham's distance) tend to be rare, which is consistent with the neutral theory of molecular evolution. However, nonsynonymous substitutions involving Grantham's distances of 58 and 61 are particularly frequent, with the frequencies equal to 0.16 and 0.11, respectively, much more frequent than those with smaller Grantham's distance. Is this evidence against the neutral theory of molecular evolution?

Part II:

Grantham's D	Frequency	Proportion
5	1	0.0076
10	8	0.0611
14	7	0.0534
21	1	0.0076
22	9	0.0687
23	2	0.0153
26	1	0.0076
27	4	0.0305
29	4	0.0305
30	8	0.0611
32	3	0.0229
38	1	0.0076
41	0	0.0000
46	2	0.0153
50	0	0.0000
53	2	0.0153
54	0	0.0000
55	0	0.0000
57	3	0.0229
58	21	0.1603
60	3	0.0229
61	15	0.1145
64	0	0.0000
65	4	0.0305
68	2	0.0153
71	0	0.0000
73	1	0.0076
76	0	0.0000
78	0	0.0000
81	0	0.0000
83	1	0.0076
87	0	0.0000
89	3	0.0229
91	0	0.0000
92	0	0.0000
94	5	0.0382
96	0	0.0000
98	0	0.0000
99	7	0.0534
101	0	0.0000

102	0	0.0000
103	0	0.0000
109	7	0.0534
112	0	0.0000
125	0	0.0000
126	3	0.0229
138	0	0.0000
142	0	0.0000
143	0	0.0000
144	0	0.0000
149	0	0.0000
152	0	0.0000
155	3	0.0229
158	0	0.0000
160	0	0.0000
177	0	0.0000
180	0	0.0000
184	0	0.0000
194	0	0.0000
204	0	0.0000
214	0	0.0000

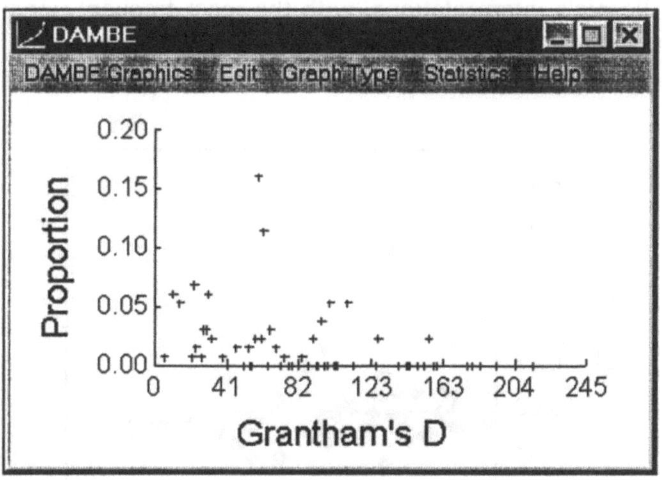

Figure 2. Observed frequencies of single-step nonsynonymous substitutions versus Grantham's distance

The relationship between the frequency of nonsynonymous substitutions and amino acid dissimilarity (measured by Grantham's distance) shown in fig. 2 is very similar to the one in Figure 1.12 in Gillespie (1991). A slightly different plot was produced by Kimura (1983) and was interpreted to mean that nonsynonymous substitutions involving similar amino acids occur more frequently than those involving different amino acids. This interpretation is consistent with the neutral theory of molecular evolution. Gillespie, however, argued that most frequently substitutions are not among those involving most similar amino acids, but those with Miyata's distance (which is another measure of amino acid dissimilarity and is positively correlated with Grantham's distance) equal to 1. This interpretation is similar to what is suggested in fig. 2, i.e., the most frequent nonsynonymous substitutions are not those involving the smallest Grantham's D, but hose with Grantham's D near 60. Can we conclude that the result favours Gillespie's interpretation against Kimura's interpretation?

We have learned in the previous chapter that nonsynonymous substitutions with different Grantham's distances are not expected to occur equally frequently, even when there is no purifying or positive selection. If we compute the expected frequencies of nonsynonymous substitutions assuming no selection (P_{exp}), and subtract P_{exp} from the observed frequencies of nonsynonymous substitution (P_{obs}), then the residuals may be attributed to the effect of selection.

P_{exp} can be obtained by clicking **Seq. Analysis|Expected P123 (Freq. Adjusted)**. The last column of the output is P_{exp}. Surprisingly, the relationship between (P_{obs} - P_{exp}) and Grantham's distance (fig. 3) still supports Gillespie's interpretations, with the most frequent nonsynonymous substitutions being those with Grantham's distances near 60.

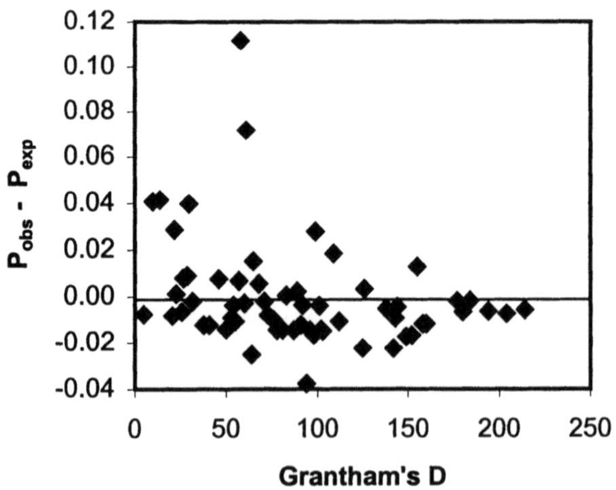

Figure 3. The most frequent nonsynonymous substitutions are not those involving the most similar amino acids, but those with Grantham's distances near 60.

Chapter 15

Case Study 4

Transition bias in mitochondrial genes of pocket gophers

1. INTRODUCTION

Transition bias in nucleotide substitution is a ubiquitous phenomenon in animal mitochondrial DNA (mtDNA), having been reported in both vertebrate species (Aquadro and Greenberg 1983; Bechkenbach et al. 1990; Brown and Simpson 1982; Brown et al. 1982; Edwards and Wilson 1990; Irwin et al. 1991; Thomas and Beckenbach 1989) and invertebrate species (DeSalle et al. 1987; Satta et al. 1987; Thomas et al. 1989; Thomas and Wilson 1991). Although the phenomenon of transition bias is poorly understood, two contributing factors have been suggested. One is that the spontaneous mutation rate involving a transitional change is much greater than that involving a transversional change (Bechkenbach et al. 1990; Brown et al. 1982; DeSalle et al. 1987; Li et al. 1984). The second is purifying selection (Li et al. 1985a), and is applicable only to protein-coding genes. Purifying selection can affect the transition bias because: 1) purifying selection tolerates synonymous mutations and eliminates nonsynonymous mutations; and 2) transitional mutations are more likely to be synonymous than transversional mutations.

The relative contribution of these two factors to the transition/transversion (s/v) ratio has not been studied in a quantitative way. We here summarize the joint effect of the two factors on the s/v ratio as follows:

$$\frac{s}{v} = \frac{\mu_s \bullet P_s}{\mu_v \bullet P_v} \tag{15.1}$$

where μ_s and μ_v are the mutation rate of transitions and transversions, respectively, and P_s and P_v are the fixation probability of a transitional mutation and a transversional mutation, respectively. Thus, transition bias can arise either from differential mutation pressure favouring transitions (i.e., a large μ_s/μ_v ratio), or from differential purifying selection against transversions, which would decrease P_v and consequently increase the P_s/P_v ratio.

At four-fold degenerate sites, both transitions and transversions are synonymous and may be assumed to be nearly neutral, with $P_{s_4} \approx P_{v_4}$, where the subscript 4 denotes four-fold degenerate sites. This leads to

$$\frac{s_4}{v_4} = \frac{\mu_s \bullet P_{s_4}}{\mu_v \bullet P_{v_4}} \approx \frac{\mu_s}{\mu_v} \tag{15.2}$$

Similarly, if we assume that purifying selection acts roughly equally against nonsynonymous transitions and nonsynonymous transversions, then $P_{s_0} \approx P_{v_0}$ (where the subscript 0 denotes nondegenerate sites), so that

$$\frac{s_0}{v_0} = \frac{\mu_s \bullet P_{s_0}}{\mu_v \bullet P_{v_0}} \approx \frac{\mu_s}{\mu_v} \tag{15.3}$$

Equations (15.2) and (15.3) state that the s_4/v_4 ratio and the s_0/v_0 ratio offer two independent estimates of the μ_s/μ_v ratio, which measures mutational contribution to transition bias. Thus, the s_4/v_4 ratio and the s_0/v_0 ratio are expected to be similar because they both reflect the same μ_s/μ_v ratio. An s/v ratio close to one at four-fold degenerate sites and at nondegenerate sites would suggest little mutational contribution to transition bias.

At two-fold degenerate sites,

$$\frac{s_2}{v_2} = \frac{\mu_s \bullet P_{s_2}}{\mu_v \bullet P_{v_2}} \tag{15.4}$$

where the subscript 2 denotes two-fold degenerate sites. Because transitions are synonymous and transversions are nonsynonymous at two-fold degenerate sites in animal mitochondrial genes, P_{s_2} is expected to be larger than P_{v_2} under neutral theory (Kimura 1983), so that the P_{s_2}/P_{v_2} ratio should be larger than one for functional genes. This P_{s_2}/P_{v_2} ratio can serve as a measure of the contribution of purifying selection to transition bias. The s/v ratio at two-fold degenerate sites is expected to increase with increasing intensity of purifying selection. An s/v ratio at two-fold degenerate sites similar to that at nondegenerate and four-fold degenerate sites suggests the absence of purifying selection.

The intensity of purifying selection against nonsynonymous substitutions can be assessed by the following three ratios:

$$\frac{P_{s_2}}{P_{v_2}} = \frac{s_2 \mu_v}{v_2 \mu_s} \tag{15.5}$$

$$\frac{P_{s_4}}{P_{s_0}} = \frac{s_4}{s_0} \tag{15.6}$$

$$\frac{P_{v_4}}{P_{v_0}} = \frac{v_4}{v_0} \tag{15.7}$$

These three ratios are expected to be the same if we assume that $P_{s_0} = P_{v_0} = P_{v_2}$ (i.e., all nonsynonymous mutations are subject to equally intense purifying selection and have the same probability of fixation regardless of whether they occur at nondegenerate or two-fold degenerate sites), and $P_{s_2} = P_{s_4} = P_{v_4}$ (i.e., all synonymous mutations are nearly neutral and have the same probability of fixation regardless of whether they occur at two-fold degenerate or four-fold degenerate sites). These assumptions have never been critically examined, although they are generally accepted as true when calculating the rate of synonymous and nonsynonymous substitutions (Li 1993; Li et al. 1985a; Nei and Gojobori 1986).

It is possible for the first assumption ($P_{s_0} = P_{v_0} = P_{v_2}$) to be violated because, when a nonsynonymous mutation occurs, the original and the replacement amino acids could be very similar to each other in physical and chemical properties, or they could be very different. If nonsynonymous mutations at two-fold degenerate sites tend to involve amino acid pairs that are more similar to (or more different from) each other than do nonsynonymous mutations at nondegenerate sites, then P_{v_2} would be larger (or smaller) than either P_{0_s} or P_{0_v}, so that the P_{s_2}/P_{v_2} ratio would be smaller (or larger) than the other two ratios. This could bias estimates of the rate of synonymous and nonsynonymous substitutions.

The three ratios in equations (15.5)-(15.7) can be used to study differential purifying selection acting on different genes in the same genome. The three ratios can all be considered as measures of the strength of purifying selection, with stronger purifying selection being correlated with larger ratios.

There are at least two more reasons for a detailed study of the relative contribution of mutation and purifying selection to transition bias. First, if purifying selection is a dominant factor shaping the rate of nucleotide substitution, then about 72% of the nucleotide sites (i.e., the proportion of nonsynonymous sites) are constrained. Such a large proportion of constrained sites would bring into question the concept of the molecular clock, because the presence of such a clock would now depend largely on the constancy of purifying selection. At present, there is little evidence that purifying selection is constant over geological time.

An understanding of the relative contribution of mutation and purifying selection to transition bias would also help us to choose phylogenetic method for systematic analysis. For example, certain computer programs such as DNAML and DNADIST in the PHYLIP package (Felsenstein 1993) include a correction for transition bias by allowing the user to input a single s/v ratio. Such implementation would be adequate if mutation is the dominant factor shaping the transition bias, but would be insufficient if there is strong purifying selection generating great heterogeneity in the s/v ratio at nondegenerate, two-fold degenerate, and four-fold degenerate sites. For nuclear genes, this heterogeneity appears to be small, with s/v ratios equal to ~4 at two-fold degenerate sites and ~2 at nondegenerate and four-fold degenerate sites (Li et al. 1985a). How great the heterogeneity is in mitochondrial genes is unknown. Considering that the ratio of synonymous to nonsynonymous substitutions is much greater in mitochondrial genes than in nuclear genes (Thomas and Beckenbach 1989), we suspect that the effect of purifying selection is greater in mitochondrial genes than in nuclear genes, which would lead to greater heterogeneity in the s/v ratio at nondegenerate, two-fold degenerate, and four-fold degenerate sites.

In this case study, we will investigate the relative contribution of mutation and purifying selection to transition bias by using mtDNA sequence data for the cytochrome-b gene (cyt-b) in 15 species of pocket gophers (Rodentia: Geomyidae). The published study (Xia et al. 1996) also includes the cytochrome oxidase subunit I (COI) gene. The results from the two genes are similar and I choose only one of them to save space. We ask the following questions: (1) Is the s/v ratio at nondegenerate sites similar to that at four-fold degenerate sites as expected? (2) Is the s/v ratio at two-fold degenerate sites greater than that at four-fold degenerate and nondegenerate sites? (3) Are the three ratios in equations (15.5)-(15.7) similar to each other?

The original paper, which includes both the COI and the cyt-b genes, also asked two additional questions: 1) Are the three ratios larger for the COI gene than for the cyt-b gene? and 2) Has COI experienced stronger purifying selection than cyt-b in the evolution of the 15 species of pocket gophers? Interested reader may consult the original paper for answers to these questions.

2. GET SEQUENCE DATA

We study mtDNA from the cyt-b gene in 15 species of pocket gopher representing six genera. These sequences (402 bp) have been deposited in GenBank with accession numbers of L11900, L11902, L11906 and L38465-L38476. It is easy to get the DNA sequences once we have GenBank accession numbers. Make sure that your computer is connected to internet either by a network card, or by a modem via PPP or SLIP. Start DAMBE and click **File|Read sequences from GenBank**. A dialog box appears (fig. 1). Type in the accession numbers separated by a comma into the text box, choose FASTA format in the dropdown box, and click the **Retrieve** button. The reason for using the FASTA format is that the retrieval can be much faster because sequences in the FASTA format do not contain detailed sequence description seen in the GenBank format.

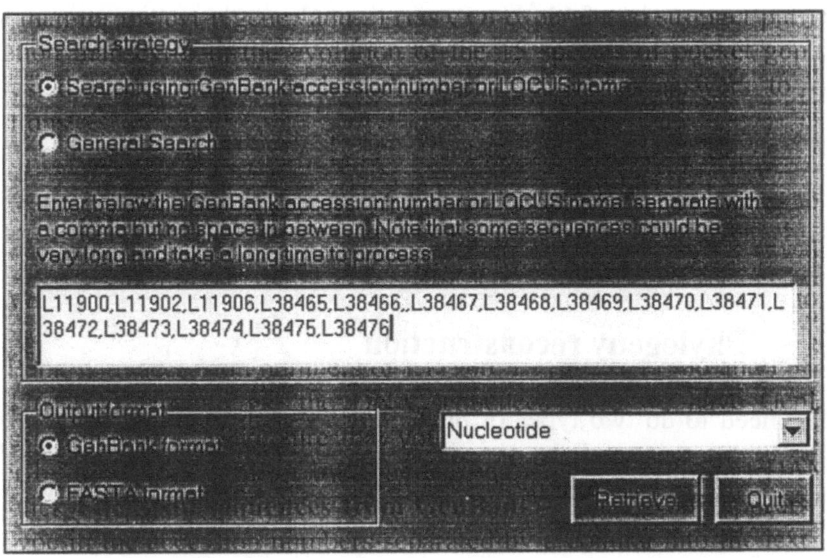

Figure 1. Retrieving the cyt-b sequences from GenBank

Once DAMBE has retrieved sequences, and if sequences are not aligned, then you will be asked whether to align the sequences or not. If you click **Yes**, then DAMBE will align the nucleotide sequences by using ClustalW. However, because the nucleotide sequences that we have retrieved are from protein-coding genes, it is better to first translate the nucleotide sequences into amino acid sequences, align the amino acid sequences, and then align the nucleotide sequences against the aligned amino acid sequences. This procedure is described in detail in the chapter dealing with sequence alignment. Save the aligned sequences to a file.

3. DATA ANALYSIS

Published estimates of the s/v ratio are typically based on pair-wide comparisons. For example, with 15 DNA sequences from pocket gopher species, we could make 105 pair-wise comparisons and report the average. However, there are two major disadvantages with this type of pair-wise comparisons in studying transition bias. First, the estimates are not statistically independent. For example, if there is one species that has recently experienced a large number of transitions and few transversions, then all 14 pair-wise comparisons between this species and the other 14 species will each contribute one data point with a large transition bias. Second, one has to assume that the nondegenerate, two-fold degenerate, and four-fold degenerate sites have remained nondegenerate, two-fold degenerate, and four-fold degenerate throughout the entire evolutionary history of the species studied. This is a tenuous assumption because the three categories of sites could potentially change into each other (i.e., a nondegenerate site could become a two-fold degenerate site, which in turn could become a four-fold degenerate site). One way to avoid these problems is to reconstruct ancestral states of DNA sequences and estimate the number of transitions and transversions between neighbouring nodes on the phylogenetic tree. Thus, first of all, we need a tree.

3.1 Phylogeny reconstruction

We need to do two types of phylogenetic reconstruction. The first is to find the best tree, and the second is to use the best tree to reconstruct ancestral sequences. Both are rather complicated processes and you should get ready for a rough ride.

3.1.1 Finding the best tree

We have not yet dealt with the conceptual issues of phylogenetic reconstruction. However, you have already leaned how to produce a tree in the case study dealing with Elongation Factor-1α. It is good for you to have more exposure to phylogenetic methods implemented in DAMBE at this point, in spite of the fact that you have not yet been formally introduced to molecular phylogenetics. This will help you appreciate the great potential of applying phylogenetic methods. For the time being, just remember that there are three major categories of phylogenetic reconstruction: the distance method, the maximum parsimony method, and the maximum likelihood method. What we are going to do now is to use the maximum parsimony method to reconstruct a tree, and then use this tree to reconstruct ancestral states using the maximum likelihood method.

If DAMBE is still active, and if the aligned sequences are still in DAMBE's buffer, then just click **Phylogenetics|Maximum parsimony|DNAMP.** A dialog box appears (fig. 2) for you to choose options used in the maximum parsimony reconstruction. Use one of the two Thomomys species as the outgroup and leave everything else untouched. Click the **Go!** button to do a phylogenetic reconstruction using the parsimony criterion.

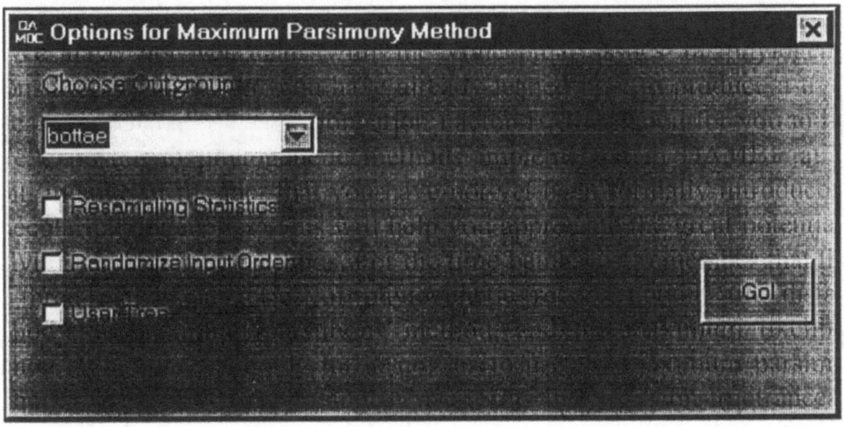

Figure 2. Dialog for phylogenetic reconstruction using the maximum parsimony method

DAMBE will find three most parsimonious trees for the cyt-b gene. You will be asked to save trees to a file, with the file extension **.nhm**, which stands for **m**ultiple trees in PHYLIP format. DAMBE's convention is to use the file extension **.dnd** for tree files with just a single tree, but **.nhm** for tree files with multiple trees. The equally parsimonious trees will be plotted on the screen. You may just close the tree window after viewing it, although it is harmless to click some menu items to do some exploration of tree manipulation.

It is troublesome to have multiple trees that are equally good (parsimonious), because now you do not know how to proceed. You need just one tree for reconstructing ancestral sequences and for pair-wise comparisons between neighboring nodes along the tree, but now you are presented with several trees. Which tree should you choose?

One way to proceed is to use the maximum likelihood method to evaluate these three most parsimonious trees and see which one has the largest likelihood. Click **Phylogenetics|Maximum likelihood|Nucleotide sequences**. A dialog box appears for you to choose options (fig. 3). Note that the default **Run mode** is **Search for best tree**. Click the dropdown arrow of the **Run Mode** combo box and choose **User tree**. This is the option that we can use to evaluate relative statistical support of alternative hypotheses.

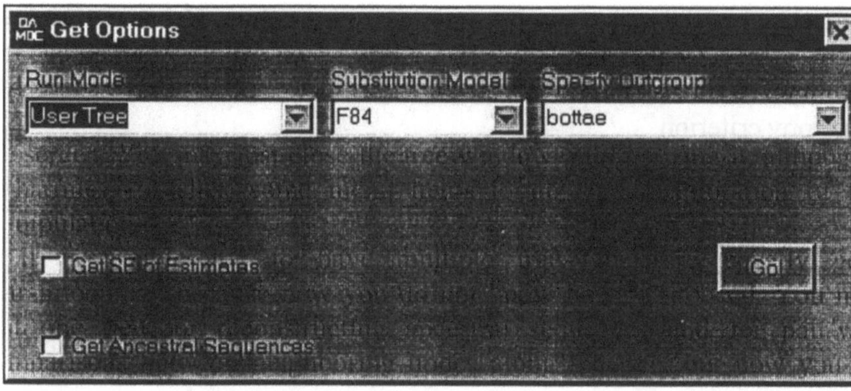

Figure 3. Evaluate the three alternative topologies using the maximum likelihood method

Click the **Go!** button and wait for a few minutes. You will be asked if you wish to evaluate relative statistical support. Click **Yes**. Wait for another minute or two, and the result will be displayed in DAMBE's display window. The following is at the end of the output:

```
Prob(i)-the probability of tree i being as good as
        the best tree, i.e., tree 1

k(i)    -the number of ML values falling with the range
        between ML(i) and MaxML, including ML(i) and MaxML.

============================================================
```

```
Tree ML         ML-MaxML  SE_Diff  RELL(i)  q(i)    k  Prob(i)
 1  -2,565.67 <====Best tree.
 2  -2,566.69 -1.0166   4.3047   0.2774  0.2362  3   > 0.5
 3  -2,566.62 -0.9536   4.3289   0.2930  0.2203  2   > 0.5
```

The output might appear confusing to you at this moment because we have not yet covered any statistical test of phylogenetic hypotheses. For the time being, just note that Tree 1 is the best tree and will be used for phylogenetic reconstruction of ancestral sequences. Also note that the best tree is not significantly better than the other two alternative trees.

It is always a good practice to have a look at the trees before carrying out further analysis. In the display window, find Tree #1 and its text representation, shown in bold below. Select the tree so that it will be highlighted, and click **Edit|Copy**. Then click **File|Paste tree into tree display panel**, and the tree will be displayed in DAMBE's tree-viewing window (fig. 4). The tree makes sense to me, so we will save it to a tree file for reconstructing ancestral sequences, which requires a file with a single tree. Click **Tree|Save tree for future viewing**, and a standard **File/Save** dialog box appears. Type in **cytb-anc.dnd** as a file name. The file type **.dnd** is short for dendrogram and is DAMBE's choice for files with a single tree.

```
Tree # 1 ...
......
((((((cavator:0.04648,((heterodu:0.02268,cherriei:0.02392):
0.01702,underwoo:0.05117):0.00668):0.02184,hispidus:
0.02786):0.07339,trichop:0.08160):0.01374,(((merriam:
0.05236,castano:0.05007):0.02725,bulleri:0.07113):0.02099,
(brevicep:0.07205,((burma:0.06466,burha:0.03457):0.03836,
personat:0.04260):0.03026):0.03634):0.02141):0.04464,
T._talpoid:0.10312):0.05413, bottae:0.05413);
```

An alternative of saving the tree is to simply select the tree in the output above, so that the bolded text will be highlighted in the display window. Now click **File|Save display buffer**. DAMBE will notice that you have selected a tree and will pop up a **File/Save** dialog box to save the selected tree. If there is only one tree is selected, then the default file type is **.dnd**. If consecutive trees are selected, then the default file type is **.nhm**.

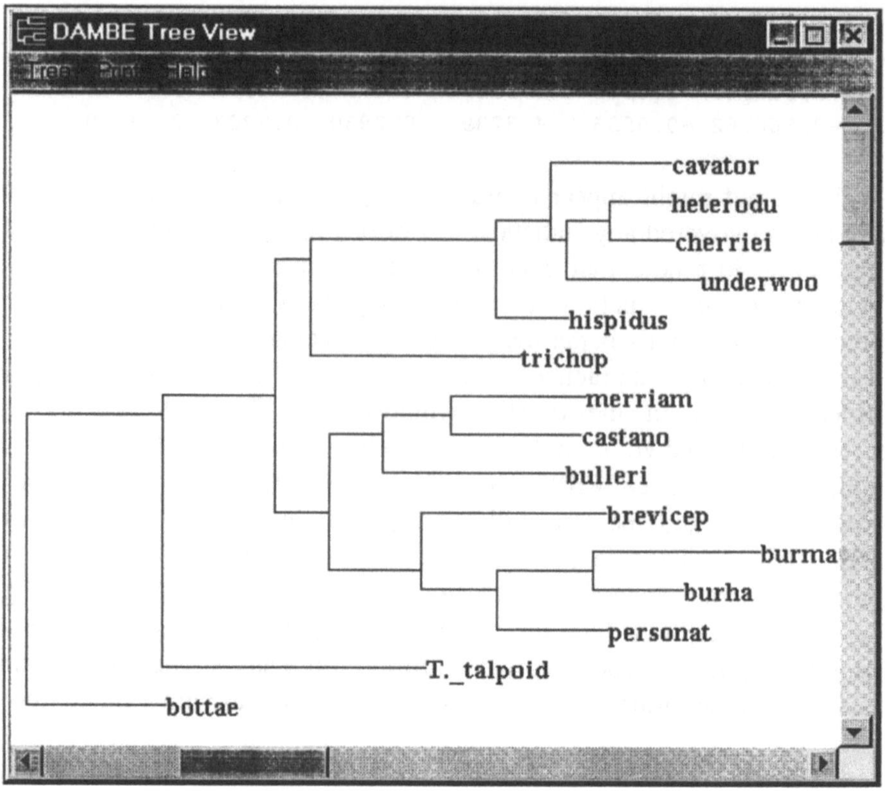

Figure 4. Phylogenetic tree used for constructing ancestral sequences

3.1.2 Reconstructing ancestral sequences

We will now use the saved tree, named **cytb-anc.dnd,** to reconstruct ancestral sequences using the maximum likelihood method. Click **Phylogenetics|Maximum likelihood|Nucleotide sequences**. A dialog box appears. Click the **Run mode** dropdown menu and change the default to **User tree**. Check the **Get ancestral sequences** checkbox (fig. 5). Finally, click the **Go!** button to begin reconstruction, which will take quite a long while. If you wish to use the computer for other purposes, such as word processing using Microsoft WORD, remember to open WORD before clicking the **Go!** button. You cannot open another program once the reconstruction process is initiated.

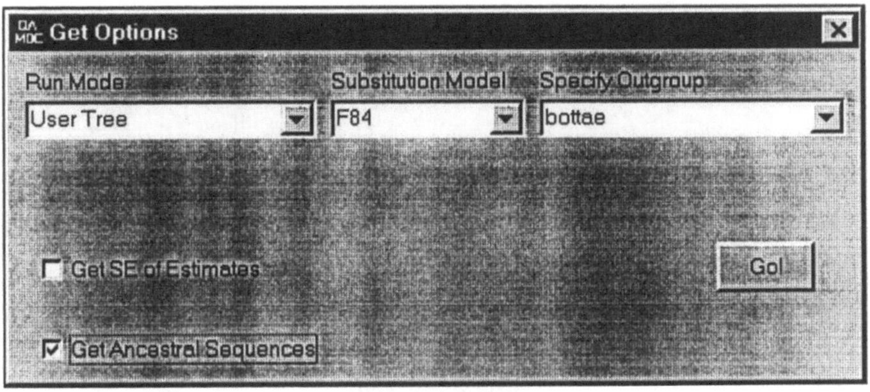

Figure 5. DAMBE's dialog box for reconstructing ancestral sequences

Once the reconstruction process is finished, you will be prompted to save the reconstructed file. The reconstructed ancestral sequences, together with the original sequences and the tree topology used for reconstruction, are saved in what I call the RST format earmarked in DAMBE by the **.rst** file extension. All the sequences and the tree are now ready for you to do pair-wise comparisons between neighboring nodes along the tree. Note that the ancestral species are named Node#i, where i is a number. If you have N sequences, then i will start from N+1.

3.2 Pair-wise comparisons between neighboring nodes

Recall that our objective is to obtain the transition/transversion (s/v) ratio at the nondegenerate, two-fold degenerate, and four-fold degenerate sites in equations(15.1)-(15.7). This is essentially a codon-based analysis, with the counting procedure detailed in Li et al. (1985a) and Li (1993). Click **Seq. Analysis|Codon differences|Li93 detail**. You will be asked to let DAMBE carry out site-wise deletion of unresolved codons. This is necessary to carry out the codon-based analysis, so click **Yes**, otherwise the operation will be aborted. A dialog box (fig. 6) with two lists appears for you to select sequence pairs for analysis. Note that the sequence pairs shown in the left listbox are not all possible pairs, but instead are neighboring nodes along the phylogenetic tree. Click the **Add all** button, and then the **Go!** button.

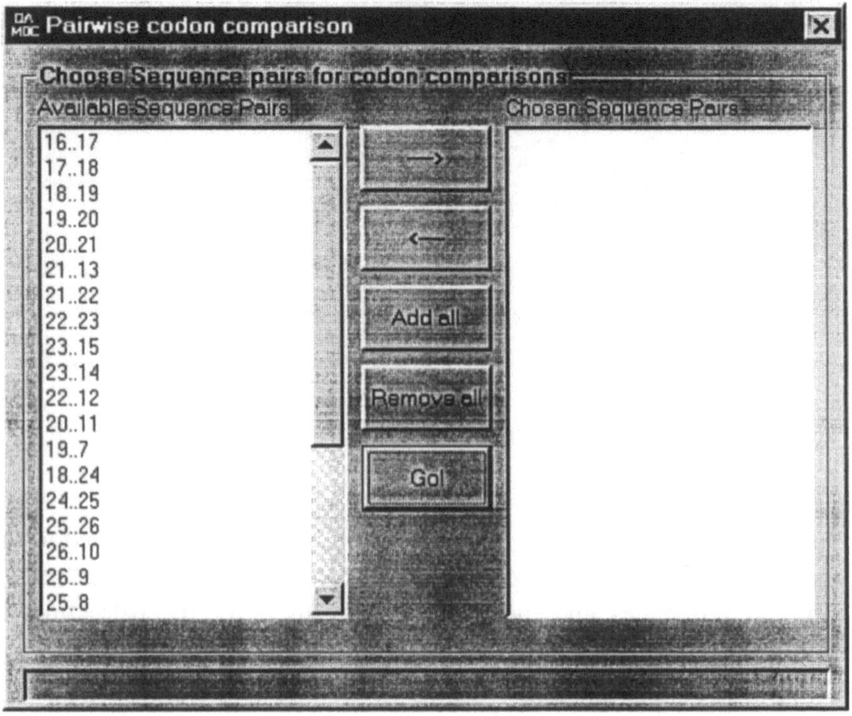

Figure 6. Choose sequences to perform codon-based analysis

You will be prompted for a file name to save the result. Just type in any file name and click the **Save** button. The result will be saved and a partial output for a single pair comparison is reproduced below:

```
Data for illustrating Li93's method

For each pairwise comparisons, the 0-fold, 2-fold and
4-fold degenerate sites are counted for each sequences.
The average of the two sequences is reported as Sites.

The two sequences are then compared codon by codon, and
the number of transitions and transversions at each of
the three categories of sites are counted, and reported
as Ns, Nv.

Kimura's two-parameter method is then applied to correct
for multiple hits, yielding the corrected number of
transitions and transversions as A and B.
```

```
Estimated synonymous and nonsynonymous substitutions
according to the Li93 method are shown as Ks and Ka,
together with their respective standard error.

node#16    - node#17
--------------------------------------------
i-fold:                 0          2          4
Num. Sites          249.5       86.5       63.0
Ns(i)             1.00000   17.50000    8.50000
Nv(i)             0.00000    0.00000    7.00000
A(i)              0.00402    0.25928    0.17696
B(i)              0.00000    0.00000    0.12566
Ks =  0.35025 SE(Ks)  =  0.07114
Ka =  0.00402 KE(Ka)  =  0.00403
```

The **i-fold** in the output with three categories 0, 2 and 4 designates the nondegenerate, two-fold degenerate and four-fold degenerate sites. **Num. Sites** shows the actual counted sites averaged between the two sequences. **Ns, Nv, A(i)** and **B(i)** have already been explained in the output. We are interested only in the estimated number of transitions and transversions at the nondegenerate, two-fold degenerate and four-fold degenerate sites, i.e., A(0), A(2), A(4) and B(0), B(2) and B(4), which have been shown in bold above. These estimates were shown in the first line row of data in Table 1.

Table 1. The estimated number of transition and transversion substitutions at nondegenerate, two-fold and four-fold degenerate sites.

Species Pair	0	2	4	0	2	4
	s_0	s_2	s_4	v_0	v_2	v_4
Node#16-node#17	0.0040	0.2593	0.1770	0.0000	0.0000	0.1257
Node#16-bottae	0.0000	0.0000	0.0218	0.0000	0.0000	0.0435
Node#17-node#18	0.0034	0.0755	0.0397	0.0068	0.0176	0.0795
Node#17-T._talpoid	0.0027	0.2244	0.1237	0.0054	0.0000	0.2474
Node#18-node#19	0.0040	0.0240	0.0052	0.0000	0.0000	0.0104
Node#18-node#24	0.0000	0.0362	0.0499	0.0000	0.0000	0.0480
Node#19-node#20	0.0040	0.2362	0.0579	0.0000	0.0000	0.1158
Node#19-trichop	0.0047	0.1493	0.1150	0.0094	0.0091	0.1003
Node#20-node#21	0.0040	0.0491	0.0330	0.0000	0.0000	0.0160
Node#20-hispidus	0.0000	0.0494	0.0697	0.0000	0.0000	0.0492
Node#21-cavator	0.0000	0.0761	0.1136	0.0000	0.0118	0.0685
Node#21-node#22	0.0013	0.0117	0.0000	0.0027	0.0000	0.0000
Node#22-node#23	0.0000	0.0235	0.0517	0.0000	0.0000	0.0164
Node#22-underwoo	0.0040	0.1167	0.1335	0.0000	0.0000	0.0505
Node#23-heterodu	0.0040	0.0357	0.0880	0.0000	0.0000	0.0000
Node#23-cherriei	0.0081	0.0362	0.0500	0.0040	0.0000	0.0000
Node#24-node#25	0.0000	0.0362	0.0320	0.0000	0.0000	0.0313
Node#24-node#27	0.0041	0.0491	0.0680	0.0000	0.0000	0.0646
Node#25-node#26	0.0014	0.0747	0.0322	0.0027	0.0000	0.0156

Species Pair	0	2	4	0	2	4
Node#25-bulleri	0.0000	0.2193	0.0895	0.0000	0.0000	0.1012
Node#26-merriam	0.0041	0.0770	0.0514	0.0061	0.0179	0.0842
Node#26-castano	0.0013	0.1453	0.0533	0.0027	0.0146	0.0640
Node#27-brevicep	0.0040	0.1897	0.1015	0.0000	0.0000	0.1213
Node#27-node#28	0.0000	0.0371	0.0675	0.0000	0.0000	0.0641
Node#28-node#29	0.0040	0.0924	0.0665	0.0000	0.0000	0.0313
Node#28-personat	0.0081	0.1206	0.0835	0.0000	0.0000	0.0154
Node#29-burma	0.0000	0.1846	0.2002	0.0000	0.0000	0.0473
Node#29-burha	0.0000	0.0638	0.0503	0.0000	0.0000	0.0484
Sum	0.0713	2.6927	2.0255	0.0397	0.0709	1.6596

There are altogether 28 such pair-wise comparisons along the phylogenetic tree, each taking up one data line in Table 1. Note that the substitution rate is high at the four-fold degenerate sites, and low at the nondegenerate sites.

The maximum likelihood estimate of the s/v ratio is

$$\frac{s}{v}=\frac{\sum_{i=1}^{N} s_i}{\sum_{i=1}^{N} v_i} \qquad (15.8)$$

where N is the number of branches, and s_i and v_i are the estimated number of transitions and transversions, respectively, between two neighbouring nodes of the *i*th branch. For example, the s/v ratio at two-fold degenerate sites is

$$\frac{s}{v}=\frac{2.6927}{0.0709}=37.97 \qquad (15.9)$$

4. RESULTS

The s/v ratio at nondegenerate sites is similar to that at four-fold degenerate sites for the cyt-b gene (Table 2), which is consistent with expectations based on equations (15.2)-(15.3). That is, both are estimates of the same parameter, i.e., mutation bias (μ_s/μ_v). The s/v ratio at two-fold degenerate sites is much greater (Table 2), suggesting that nucleotide substitution at the two-fold degenerate sites is constrained by strong purifying selection against transversions, which are nonsynonymous at these sites.

Table 2. The s/v ratio at the nondegenerate, two-fold degenerate and four-fold degenerate sites

Site	s	v	s/v
0	0.0713	0.0397	1.80
2	2.6928	0.0709	37.98
4	2.0255	1.6596	1.22

The high rate of nucleotide substitutions at the four-fold degenerate sites for the cyt-b genes (Table 1-2) indicates substitutional saturation. Because substitutional saturation eventually leads to a reduction of available information for estimating the number of transitions and transversions, the s/v ratio for the four-fold degenerate sites based on all pair-wise comparisons between neighbouring nodes may be a biased estimate. To obtain an less biased estimate of the s/v ratio for the four-fold degenerate sites, we need data with negligible substitutional saturation (i.e., recently diverged taxa) and without mutual statistical dependence among data points. For this reason, you may select a subset of pair-wise comparisons involving more closely related species. The resulting s/v ratio is expected to be greater (≈ 2), and closer to that observed at the nondegenerate sites (Table 1), where little substitutional saturation should have occurred. Try it.

Our results show that the contribution of mutation (μ_s/μ_v) to the s/v ratio is relatively small, and clearly cannot explain the much larger s/v ratio observed at the two-fold degenerate sites (Table 1 and Fig. 2). Given that the μ_s/μ_v ratio estimated from the four-fold degenerate and nondegenerate sites is about two, the P_{s_2}/P_{v_2} ratio should be ~20 for cyt-b to account for the observed s/v ratio at the two-fold degenerate sites according to equation (15.4) and Table 1-2. Because transitions are synonymous and transversions nonsynonymous at two-fold degenerate sites, the observed transition bias at two-fold degenerate sites is attributable to purifying selection acting against amino acid substitutions.

Three ratios in equations (15.5)-(15.7) are estimates of the intensity of purifying selection. P_{s_2}/P_{v_2} has already been calculated as 20, given that the μ_s/μ_v ratio is around 2. The P_{v_4}/P_{v_0} ratio is also around 20 because the v_4/v_0 ratio is 41.8 from data in Table 2. The P_{s_4}/P_{s_0} ratio, however, is only about 15, which seems much smaller than the other two ratios. Do you have an explanation for this difference?

One possible explanation is simply substitution saturation. There are a lot of transversions occurring at the four-fold degenerate sites (Table 1-2). An increase in transversions will necessarily lead to a reduction in observed number of transitions. For example, suppose that two sequences (Seq1 adn Seq2) originally have a "homologous" nucleotide site occupied by A. In Seq1, a transition occurs at this site, changing A to G, which is followed by a transversion, changing G to T. When we compare Seq1 and Seq2 at this particular site, we see A on Seq2 and T on Seq1. So we score a transversion, but miss the previous A→G transition. If a subsequent transition changes the T to C at this site of Seq1, then we will observe A on Seq2 (the original nucleotide) and C on Seq1 (the nucleotide after three changes). So we will still score a transversion, in spite of the fact that two transitions have also occurred at this site. In short, increasing number of transversions will result in a decrease in observed transitions.

At the four-fold degenerate sites, many transversions have occurred, with the sum of transversions being 1.6596 (Table 2), much greater than the corresponding values at the nondegenerate and two-fold degenerate sites. It is therefore not surprising that the P_{s4}/P_{s0} ratio, when estimated from the s_4/s_0 ratio, should be smaller than the P_{s2}/P_{v2} ratio or the P_{v4}/P_{v0} ratio because s_4 is severely underestimated due to the large number of transversions that have occurred at the four-fold degenerate sites.

5. DISCUSSION

Our results suggest that transition bias in protein-coding sequences of mtDNA is mainly caused by strong purifying selection acting against transversions at two-fold degenerate sites. However, differential mutation pressure also may have contributed to the transition bias. The s/v ratio at nondegenerate and four-fold degenerate sites is approximately 2 (Table 1), which suggests a higher spontaneous rate of transitional mutations relative to transversional mutations in these species of pocket gophers. Whether our conclusions are generally applicable requires further empirical studies.

The s/v ratio at two-fold degenerate sites in the two mitochondrial genes studied is much greater (= 40) than that reported for mammalian nuclear genes, where the s/v ratio at these sites is about 4 (Li et al. 1985a). However, the s/v ratios at nondegenerate and four-fold degenerate sites for our mitochondrial genes (about 2) are comparable to those for nuclear genes (Li et al. 1985a). This suggests that the mutational contribution to transition bias is similar between nuclear DNA and mtDNA, and that the dramatic difference in the s/v ratio at the two-fold degenerate sites between mitochondrial genes and nuclear genes is attributable to much stronger purifying selection against nonsynonymous mutations in mitochondrial DNA than in nuclear DNA. This is corroborated by our result that the P_{s4}/P_{s0} and P_{v4}/P_{v0} ratios are also much greater for the mitochondrial gene than for the nuclear genes.

The dramatic heterogeneity in the s/v ratio among nondegenerate, two-fold degenerate and four-fold degenerate sites shown in the cyt-b gene (Tables 1) argues strongly against the common practice of lumping all transitions and all transversions together to obtain an overall s/v ratio. Such a ratio is of little meaning, and it obscures important biological information.

A substitutional process is characterized by the rate of transition and the rate of transversion, and it is evident that substitutional processes at nondegenerate, two-fold degenerate, and four-fold degenerate sites differ. For example, the substitutional process at nondegenerate sites differs from that at two-fold degenerate sites mainly in the rate of transitional substitution, and it differs from that at four-fold degenerate sites in the rate of both transitional and transversional substitutions. Finally, the process at two-fold degenerate sites differs from that at four-fold degenerate sites in the rate of transversional substitution. Thus, the overall process of nucleotide substitution in protein-coding sequences is much more complex than is assumed in current computer programs used for phylogenetic analysis. To reduce this heterogeneity in nucleotide substitution at nondegenerate, two-fold degenerate, and four-fold degenerate sites, one should either use the amino acid-based maximum likelihood method (e.g., Kishino et al. 1990) for phylogenetic reconstruction, or use the codon-based maximum likelihood method (Goldman and Yang 1994).

Chapter 16

Substitution Pattern in Amino Acid Sequences

DAMBE offers two ways to look at the substitution pattern of amino acid sequences. One is more suitable for sequences in the RST format, which contains the sequences from extant OTUs, a phylogenetic tree, and the ancestral sequences reconstructed from the extant sequences based on the phylogenetic tree. Pair-wise sequence comparisons are made only between neighboring nodes along the tree. The other output is more appropriate for all other sequence formats without a tree structure and ancestral sequences, and the output is an empirical substitution matrix from all pair-wise comparisons.

1. SUBSTITUTION PATTERN FROM SEQUENCES IN RST FORMAT

Start DAMBE and open a file containing amino acid sequences and reconstructed ancestral sequences. If you do not have such a file, then just open the **invert.rst** file that comes with DAMBE and translate the nucleotide sequences into amino acid sequences by click **Sequences|Work on amino acid sequence**. Now click **Seq. Analysis|Amino Acid Difference**. A dialog box (fig. 1) will appear for you to specify which sequence pairs you wish to study. Note that the **invert.rst** file contains seven original sequences, labelled as 1 to 7, and five reconstructed ancestral sequences, labelled 8 to 12. The sequence pairs shown in the left list sequences on the two ends of each branch of the phylogenetic tree. Typically you would want to study all these sequences pairs, so just click the **Add all** bottom to move all the sequence pairs to the right. There are two option buttons for you to specify output format, either in column format or matrix format. The latter is meaningful only for sequences not in the RST format. The default is the column format when the input file is in RST format, and you should leave it as is.

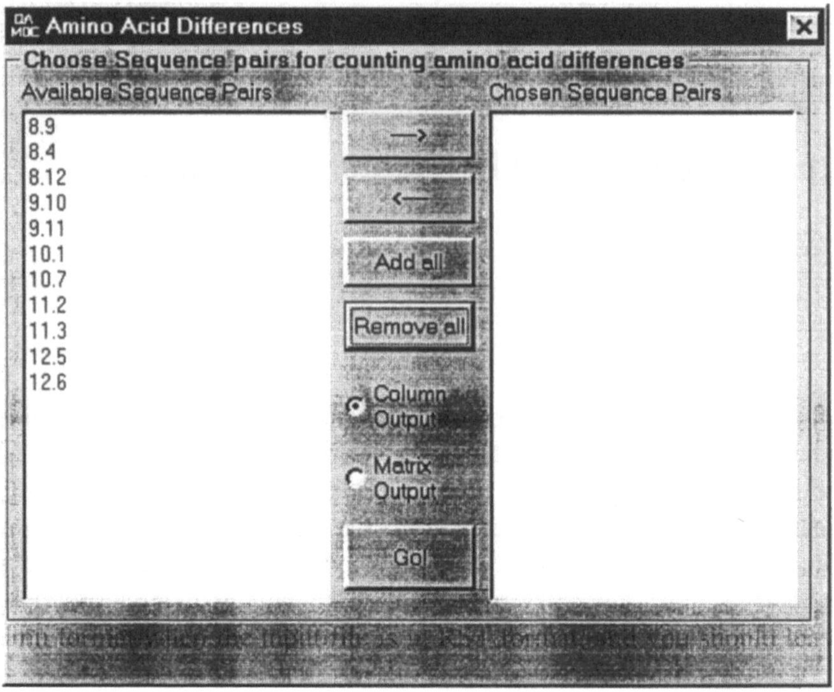

Figure 1. Choosing amino acid sequences to obtain the substitution pattern

Click the **Go!** button. A standard **File/Save** dialog box will appear for you to enter a file name for saving the output. The column output is in two parts, and is partially reproduced below:

```
Part I:
Output from pairwise comparisons for sequences in file:
D:\MS\DAMBE\inv7.RST
Dg - Grantham's distance
F(obs) - Observed number of substitutions
F(exp) - Expected number of substitutions based on amino acid
         frequencies only

AA1 AA2 F(obs)   Dg
=========================

(node#8 vs. node#9)

Gln Asp       1   61
```

```
Lys Gln      1   53
Ser Arg      1  109
Ser Pro      1   73
Thr Ala      1   58
Val Leu      1   32

(node#8 vs. Che90052)

Arg Ala      1  111
Asp Asn      1   23
Gln Ala      1   91
......
```

Part II:

```
Pooled output:

AA1 AA2   Dg   F(obs)   F(exp) Fobs-Fexp
=========================================
Arg Ala  111        2     1.53      0.47
Asn Ala  110        6     1.57      4.43
Asn Arg   85        0     1.01     -1.01
Asp Ala  126        4     1.67      2.33
Asp Arg   96        0     1.11     -1.11
......
Val Pro   68        0     1.78     -1.78
Val Ser  123        2     1.56      0.44
Val Thr   69        2     1.80      0.20
Val Trp   88        0     1.13     -1.13
Val Tyr   55        0     1.31     -1.31
=========================================
```

The first part consists of detailed information on amino acid substitution for individual pair-wise comparisons, e.g., the two ancestral sequences, **node #8** and **node #9**, differ by six amino acid sites. Grantham's (1974) distance is also listed. Most amino acid substitutions should be between similar amino acids with small Grantham's distances.

The second part presents pooled results from all individual pair-wise comparisons. The neutral theory of molecular evolution expects the substitution rate to decrease with increasing amino acid dissimilarity. If Grantham's distance is a good measure of amino acid dissimilarity, then the rate of substitution should decrease with increasing Grantham's distance. One might therefore be eager to see a plot of F(obs) versus Grantham's distance. However, such a plot may be misleading because there is no adjustment for amino acid frequencies. For example, the amino acid pair with the smallest Grantham's distance (= 5) is between leucine and isoleucine. If the amino acid sequences do not have leucine or isoleucine, then obviously there will be no substitution between leucine or isoleucine, and the number of observed substitutions for Grantham's distance equal to five will be zero. This might lead you to think that substitutions between the most similar amino acids are very unlikely.

The column headed by **F(exp)** lists the expected number of substitutions based on amino acid frequencies only. If there is no leucine and isoleucine in the sequences, then F(exp) will be zero. If there are many leucine and isoleucine residues in the sequences, then F(exp) will be large. The column headed by **Fobs. - Fexp.** is the difference between the observed and the expected number of substitutions. If Grantham's distance is small, then the difference should large, otherwise the difference will be small or negative. A plot of (Fobs - Fexp) versus Grantham's distance is shown in fig. 2.

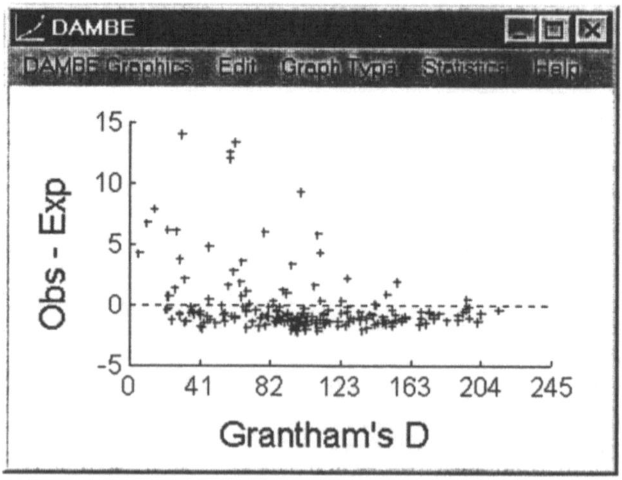

Figure 2. The relationship between the number of amino acid substitutions and Grantham's distance

Note that most substitutions are between similar amino acids (i.e., with a small Grantham's distance between them) and few are between very different amino acids (fig. 2). Kimura (1983) and Gillespie (Gillespie 1991) have each produced a similar plot, but offered different interpretations. Kimura claimed that this is consistent with the neutral theory of molecular evolution which predicted that most substitutions should be neutral and few should have a large effect (i.e., a large Grantham's distance). Gillespie, however, claimed that most frequent substitutions are not between the most similar amino acids, but between those that have a Miyata's (Miyata et al. 1979) near 1 (which roughly correspond to Grantham's distance near 70-80). With the huge amount of data available now in molecular databases, it should now be easy to discriminate between these two alternative interpretations.

2. SUBSTITUTION PATTERN FROM ALL PAIR-WISE COMPARISONS

If your input file is not in RST format, then the default output is the matrix output rather than the column output as shown in fig. 1. To see how it looks like, just open a file with amino acid sequences. If you do not have such a file, then just open the **invert.fas** file that comes with DAMBE, and translate the nucleotide sequences into amino acid sequences by click **Sequences|Work on amino acid sequence**. Now click **Seq. Analysis|Amino Acid Difference**. A dialog box (fig. 1) appears. Note that the listbox on the left side now lists all possible pair-wise comparisons. Also note that the default output is now **Matrix output** rather than **Column output**. Click the **Add all** button to move all sequence pairs to the right.

Now click the **Go!** button. You will again be asked to enter a file name for saving the result, which is partially displayed below:

```
Output from pairwise comparisons for sequences in file:
D:\MS\DAMBE\Invert5.FAS

(Chelicerl vs. Branchipod)

    Ala Arg Asn Asp ......
Ala  20
Arg   1  13
Asn   1   1  11
Asp   1   0   1  16
Cys   1   0   0   0   5
Gln   0   0   1   2   0  22
Glu   0   0   0   1   0   2   6
......
```

This output allows you to quickly identify which amino acid pair has experienced most substitutions. However, the information is much more limited than the column output. This function is mainly for documenting empirical substitution patterns when you have a lot of amino acid sequences.

should now be easy to discriminate between these two alternative ...

SUBSTITUTION PATTERN FROM ALL THE ...

Chapter 17

A Statistical Digression

1. INTRODUCTION

Although this book is intended for readers with little mathematical or statistical background, it is important at this point for the reader to gain some familiarity with some basic concepts of statistics such as probability distribution, statistical estimation and hypothesis testing, in order to better appreciate subsequent chapters. You are assumed to know some basic calculus such as how to take derivative or partial derivative. Those who are already familiar with probability distributions and statistical estimation can skip this chapter.

We will review two probability distributions that you might have already been familiar with, the binomial distribution and the multinomial distribution, and use them to introduce the maximum likelihood method. Several additional probability distributions (i.e., the Poisson, the negative binomial and the gamma) for modelling the frequency distribution of substitutions over sites will be introduced in later chapters.

The maximum likelihood method has been used extensively for parameter estimation in molecular data analysis, and the likelihood ratio test features prominently in discriminating between alternative phylogenetic hypotheses. It is beneficial to understand the fundamental rationale of the maximum likelihood method and to get familiar with the relevant terminology.

2. TWO DISCRETE PROBABILITY DISTRIBUTIONS

2.1 The Binomial Distribution and the Goodness-of-fit test.

Probability distributions model the frequency distribution of various events from statistical experiments. An event is the simplest outcome in a statistical experiment such as tossing a coin. The coin-tossing experiment has only two simplest outcomes, head or tail. A statistical experiment, or trial, with only two possible outcomes is called a Bernoulli trial, and an experiment consisting of a series of independent and identical Bernoulli trials is call a Bernoulli process. Strictly speaking, a Bernoulli process consists of n independent and identical dichotomous trials, where p is the probability of one outcome (typically designated as "success") on a given trial and remains constant from trial to trial. The restriction that p remains constant implies that the Bernoulli process is stationary. In comparative sequence analysis, a substitution model is said to be stationary if the probability of a character state, e.g., a nucleotide A or an amino acid glycine, changing into another remains constant during the evolutionary history. All substitution models, except for those underlying the paralinear (Lake 1994) and the LogDet (Lockhart et al. 1994) distances, share the stationarity assumption.

A Bernoulli process is characterized by two parameters, the number of trials (n) and the probability of success in each trial (p). The probability of failure is often designated as q (=1-p). Note that the labelling of one of the two dichotomous outcomes as "success" and the other outcome as "failure" does not have any implication on their desirability.

If you throw the fair coin only once, then there are only two possible outcomes, and the probability distribution is simply 0.5 for success and 0.5 for failure. If you throw the coin twice and record the outcome, then there are three possible outcomes. Designate head as H and tail as T, the three outcomes are HH, HT and TT, with the associated probability distribution having three values, 0.25, 0.5 and 0.25, respectively. These are the probabilities of having 2, 1 and 0 heads (successes) in the Bernoulli process consisting of two Bernoulli trials. Note that the probabilities are obtained by the expansion of $(p + q)^2$.

The probability of observing exactly r successes out of n independent and identical Bernoulli trials is

$$P(R = r \mid n, p) = \frac{n!}{r!(n-r)!} p^r q^{n-r} \tag{17.1}$$

where R is a random variable representing the number of successes in n trials from a Bernoulli process. Note that this term is one of the terms from the expansion of $(p + q)^n$.

If n = 2 and p = 0.5, then

$$\begin{aligned} P(2 \mid n, p) &= \frac{2!}{2!(2-2)!} 0.5^2 \times 0.5^{2-2} = 0.25 \\ P(1 \mid n, p) &= 0.5 \\ P(0 \mid n, p) &= 0.25 \end{aligned} \tag{17.2}$$

The mean and the variance of the binomial distribution are easy to derive. In statistics, we use E(x) to designate the expected value of x such as the mean of x. If we take a single Bernoulli trial, then P(1) = p and P(0) = 1- p, and we have

$$\begin{aligned} \bar{R} &= E(R) = 1 \cdot p + 0 \cdot (1-p) = p \\ E(R^2) &= 1^2 \cdot p + 0^2 \cdot (1-p) = p \\ Var(R) &= E(R^2) - [E(R)]^2 = p - p^2 = p(1-p) \end{aligned} \tag{17.3}$$

For n independent Bernoulli trials, the mean of the binomial distribution is the sum of the means of the individual Bernoulli trials, and the variance is similarly the sum of variances of individual Bernoulli trials, i.e.,

$$E(R) = np \tag{17.4}$$

$$Var(R) = np(1-p) = npq \tag{17.5}$$

Let us now introduce a simple goodness-of-fit test. Suppose a fish population in a very large lake with a 1:1 sex ratio. If we take a random sample of six fish, the probability of getting 0, 1, 2, ..., 6 males in the sample, according to equation (17.1), is shown in the second column of Table 1.

Table 1. Expected frequencies based on the binomial distribution and the observed frequencies from 100 samples with six fish in each sample.

N_{male}	Probability	Expected N_{sample}	Observe N_{sample}
0	0.015625	1.5625	0
1	0.09375	9.375	3
2	0.234375	23.4375	14
3	0.3125	31.25	68
4	0.234375	23.4375	11
5	0.09375	9.375	3
6	0.015625	1.5625	1

We can test whether the random variable, i.e., the number of males in the sample, really follows the binomial distribution by taking multiple samples of six each. Suppose we have taken 100 such samples and the frequency distribution of the number of samples containing various number of males are shown in the last column of Table 1.

Note that the total number of fish sampled is 600 and the total number of males in the sample happens to be exactly 300. So our assumption of equal sex ratio appears to be valid. However, the assumption of binomial distribution clearly does not hold. Two many samples have roughly equal number of males and females and two few have extreme sex ratios in comparison with what we would expect from a binomial distribution (Table 2). Whether the deviation of the observed values from the expected values is statistically significant is tested by a chi-square goodness-of-fit test:

$$\chi^2 = \frac{\sum_{i=0}^{6}(O_i - E_i)^2}{E_i} = 64 \tag{17.6}$$

With seven groups and the assumption of equal numbers of the two sexes, the test has six degrees of freedom. If we have not assumed an equal sex ratio, but instead would estimate the proportion of males from sample data, then one more degree of freedom will be lost so that the test will have only five degrees of freedom. The resulting P is 0.0000, so we reject the null hypothesis and conclude that our random variable, i.e., the number of males in the sample, does not follow the binomial distribution.

2.2 The Multinomial Distribution

The multinomial distribution is a generalization of the binomial distribution. Consider a nucleotide sequence completely randomized by mutation. We now sample this nucleotide sequence site by site. In contrast to the coin-tossing or fish-sampling experiment where there are only two possible outcomes, now we have four outcomes represented by A, C, G, T. Suppose that, at each site, the probabilities of getting A, C, G, or T are p_1, p_2, p_3, and p_4, respectively, and these probabilities do not change from site to site. If we sample n nucleotide sites, then the probability of getting n_1 of nucleotide A, n_2 of nucleotide C, n_3 of nucleotide G and n_4 of nucleotide T, where $n = \Sigma n_i$, is given by

$$P(n_1\ to\ n_4, | n, p_1\ to\ p_4) = \frac{n!}{n_1!n_2!n_3!n_4!} p_1^{n_1} p_2^{n_2} p_3^{n_3} p_4^{n_4} \quad (17.7)$$

Note that the right-hand term results from the expansion of $(p_1 + p_2 + p_3 + p_4)^n$. This example is one special case of the so-called multinomial distribution in which the number of outcomes may be any value larger than two.

3. THE SIMPLEST PRESENTATION OF THE MAXIMUM LIKELIHOOD METHOD

The maximum likelihood method is one of the methods for parameter estimation. The random variable whose parameters are to be estimated is assumed to follow an explicit probability model. In the section on the binomial distribution, we have learned a random variable which is the number of successes in a random sample. We know that the parameter p is an important parameter in a binomial distribution. How to derive a maximum likelihood estimate of p and its variance?

Suppose we wish to estimate the proportion of males of a fish population in a large lake. If we sample one fish randomly, then the sampling constitutes one Bernoulli trial. If we take 20 fish sequentially and randomly, and get 12 male fish, what is a maximum likelihood estimate of the proportion of males in the population and the variance of the proportion?

You may readily answer that the proportion of males is simply p = 12/20 = 0.6, and the variance is Var(p) = pq/n = 0.6*0.4/20. The two estimates you have just offered happen to be maximum likelihood estimates of the two parameters. Let us see how they are derived by the maximum likelihood method.

We first need to have a likelihood function which is the probability density of this particular sampling outcome:

$$L(12\ male, 8\ female\ |\ p) = \frac{20!}{12!8!} p^{12}(1-p)^8 \quad (17.8)$$

The principle of maximum likelihood requires us to choose as our estimate the value of p that maximizes the likelihood value. To facilitate this maximization process, we first take the natural logarithm of the likelihood function, take its derivative with respect to p and set it to zero. Solving the resulting equation gives us the maximum likelihood estimate of p:

$$\ln L = A + 12\ln(p) + 8\ln(1-p)$$

$$\frac{\partial \ln L}{\partial p} = \frac{12}{p} - \frac{8}{1-p} = 0 \tag{17.9}$$

$$p = \frac{12}{20} = 0.6$$

The likelihood estimate of the variance of p is equal to

$$\begin{aligned} Var(p) &= -\frac{1}{\frac{\partial^2 \ln(L)}{\partial p^2}} = -\frac{1}{-\frac{12}{p^2} - \frac{8}{(1-p)^2}} \\ &= -\frac{1}{-\frac{12}{0.6^2} - \frac{8}{(1-0.6)^2}} = 0.012 \end{aligned} \tag{17.10}$$

which is equal to 0.6(1-0.6)/20 = pq/n, which is the variance of p for the binomial distribution that you have already suggested previously.

Suppose we sample another 16 fish and found 10 to be males. Then the likelihood function is

$$L(p) = \frac{20!}{12!8!} p^{12}(1-p)^{8} \frac{16!}{10!6!} p^{10}(1-p)^{6} \tag{17.11}$$

You can demonstrate that the maximum likelihood estimate of p from the two samples is

$$p = \frac{12+10}{20+16} \tag{17.12}$$

It might seem that we have done much calculation just to derive something that you already know. However, there are two points that I wish to make. First, an appreciation of the derivation above may help you derive your own maximum likelihood estimators of unknown parameters, or at least help you to understand maximum likelihood estimators derived by others. Second, we see that the maximum likelihood method does not incorporate any prior knowledge. If we get 20 fish with no female, the maximum likelihood method will lead us to p = 1, which we know to be wrong based on our biological knowledge, i.e., a population cannot be made of all males. This is where the Bayesian approach is more preferable.

4. BIAS IN THE MAXIMUM LIKELIHOOD METHOD

The maximum likelihood estimate can be biased, which can be demonstrated by a random variable following the normal distribution whose probability density function is determined by the mean μ and the variance σ^2:

$$f(x_i) = \frac{1}{\sqrt{2\pi\sigma^2}} e^{-\frac{(x_i-\mu)^2}{2\sigma^2}} \tag{17.13}$$

Suppose we take a random sample of size n from the population, x_1, x_2, ... x_n, what is the likelihood estimate of the mean and variance of the variable x?

The likelihood function is:

$$L(x_1, x_2, ..., x_n) = \left(\frac{1}{\sqrt{2\pi\sigma^2}} e^{-\frac{(x_1-\mu)^2}{2\sigma^2}}\right)\left(\frac{1}{\sqrt{2\pi\sigma^2}} e^{-\frac{(x_2-\mu)^2}{2\sigma^2}}\right)$$

$$\cdots\left(\frac{1}{\sqrt{2\pi\sigma^2}} e^{-\frac{(x_n-\mu)^2}{2\sigma^2}}\right) \tag{17.14}$$

$$\left(\frac{1}{\sqrt{2\pi\sigma^2}}\right)^n e^{-\sum_{i=1}^{n}\frac{(x_n-\mu)^2}{2\sigma^2}}$$

$$\ln L = n \ln\left(\frac{1}{\sqrt{2\pi\sigma^2}}\right) - \sum_{i=1}^{n} \frac{(x_n - \mu)^2}{2\sigma^2} \tag{17.15}$$

Take the partial derivative of lnL with respect to μ, set it to zero and solve the resulting equation for μ, we obtain our estimate of μ (designated by $\bar{x}$) as

$$\bar{x} = \frac{\sum_{i=1}^{n} x_i}{n} \tag{17.16}$$

We can also take the partial derivative of lnL with respect with σ^2, set it to zero and solve the resulting equation for σ^2, which gives us a biased estimate (designated by s^2) of σ^2:

$$s^2 = \frac{\sum_{i=1}^{n} (x_i - \bar{x})^2}{n} \tag{17.17}$$

The unbiased estimate should have the denominator n replaced by (n - 1). However, if n is very large, then the bias is negligible. In other words, the s in equation (17.17) converges to the unbiased estimate asymptotically as n approaches infinity.

5. EXERCISE

Given the relationship in equation (17.7) for the multinomial distribution, prove that $p_i = n_i/n$ is a maximum likelihood estimate of p_i. If we take another m sites and find the observed frequencies of A, C, G, T to be m_1, m_2, m_3, m_4, respectively, what is the maximum likelihood estimate of p_i with the data of two samples?

Chapter 18

Theoretical Background of Genetic Distances

1. INTRODUCTION

Genetic distances are measures of the genetic differences between individuals, species or higher taxa. Consequently, the most frequent application of genetic distances is in phylogenetic reconstruction by using one of the distance methods, such as the unweighted pair-group method with arithmetic mean (UPGMA), the neighbor-joining (Saitou and Nei 1987), or the more complicated Fitch-Margoliash method (Fitch and Margoliash 1967). Genetic distances are also used in conservation biology because genetic uniqueness is a major criterion for species conservation. For example, the closest relatives of the giant panda are species in the bear family, yet the giant panda has a large genetic distance between itself and all those bear species. The giant panda is therefore not only an endangered, but also genetically unique species, and should consequently be given high priority in its conservation.

Genetic distances can be estimated by using the following kinds of data: (1) nucleotide sequences which include both non-protein-coding sequences and protein-coding sequences or codon sequences, (2) amino acid sequences and (3) allele frequencies. Different genetic distances are based on different substitution patterns of the molecular sequences or genes concerned.

DAMBE can compute genetic distances by using nucleotide, amino acid or codon sequences as well as allele frequency data, and the resulting distance matrix can be used directly in phylogenetic reconstruction using various distance methods implemented in DAMBE.

Different measures of genetic distances assume different models of nucleotide, amino acid or allele substitutions in the population. This chapter provides an elementary treatment of conceptual issues underlying these genetic distances.

2. GENETIC DISTANCES FROM NUCLEOTIDE SEQUENCES

Genetic distances based on nucleotide sequences are computed after you have read in a nucleotide sequence file and then clicked **Sequence Analysis|Nucleotide Difference|Detailed Output**. Six kinds of distances are generated: the Jukes and Cantor's (1969) one-parameter distance (K_{JC69}), Kimura's (1980) two-parameter distance (K_{K80}), Tajima and Nei's (1984) distance (K_{TN84}), the distance based on the F84 model (Felsenstein 1993), Tamura and Nei's (1993) distance, and Lake's (1994) paralinear distance (d_{Lake}). For the F84 and TN93 models, the distances based on the gamma-distributed substitution rates are also computed for phylogenetic analysis using distance methods.

These estimates of genetic distances have a few common assumptions: (1) the substitutions occur independently in different lineages; (2) substitutions occur independently among sites; (3) the process of substitution is described by a time-homogeneous Markov process, and (4) except for d_{Lake}, the process of substitution is stationary. In other words, the frequencies of nucleotides have remained constant over the time period covered by the data. JC60 and K80 models also assume equal nucleotide frequencies.

A substitution model typically has two categories of parameters for describing a substitution pattern (Kumar et al. 1993; Li 1997, p. 67; Yang 1997b), i.e., the rate ratio parameters (often symbolized by α, β, δ, γ, etc.) and the frequency parameters, i.e., π_A, π_C, π_G, and π_T. It is important to realize at the very beginning that the validity of a genetic distance depends much on how close its underlying model of nucleotide substitutions approximates reality. So it may be helpful to spend just a bit more time to have a closer look at those assumptions of substitution models.

The first assumption, that substitutions occur independently in different lineages, is the least problematic, and is expected to be true excluding the following two cases. First, two different lineages may come to interbreed in a reticulate speciation event. Second, horizontal transfer of genes happens which is typically mediated by viral agents. Most people would agree that these two events are so rare in evolution that we can generally ignore their effect.

The second assumption, that substitutions occur independently among sites, is obviously not true. Take two codons, GAT and GGT, for example. Both codons end with a T. Whether a T→A substitution would occur depends much on whether the second position is an A or a G. The T→A substitution is rare when the second codon position is A because a T→A mutation in the GAT codon is nonsynonymous, but relatively frequent when the second codon position is G because such a T→A mutation in a GGT codon is synonymous. This violates the assumption that nucleotide substitutions occur independently among sites. One can easily come up with many other examples in which the assumption does not hold.

The third assumption that the process of substitution is described by a time-homogeneous Markov process is also unlikely to be true. The recent proposal of punctuated equilibrium based on fossil records suggests that evolution occurs very sporadically, rather than continuously. Limited molecular evidence also favour episodic evolution (Gillespie 1991).

The fourth assumption, that the process of substitution is stationary, is also problematic. For example, there are GC-rich and AT-rich isochores in vertebrate genomes, which are presumably maintained by differential mutation pressure, with GC-rich isochores having GC-biased mutation and AT-rich isochores having AT-biased mutations. When a gene originally located in a GC-rich isochore gets relocated to an AT-rich isochore, then we expect directional GC to AT mutations so that the relocated gene will eventually become AT-rich. This substitution process, which leads to a decrease of the frequencies of C and G and an increase in the frequencies of A and T, is clearly not stationary.

2.1 JC69 and TN84 distances

Jukes and Cantor's (1969) model assumes independent nucleotide substitutions at all sites with equal probability. Whether a base changes is independent of its identity, and when it changes there is an equal probability of ending up with each of the other three bases.

A maximum likelihood estimator for the Jukes and Cantor's distance is

$$K_{JC} = -\frac{3}{4}\ln\left(1-\frac{4p}{3}\right) \tag{18.1}$$

where p is assumed to be equal to the observed proportion of nucleotide differences between the two sequences. The large-sample variance of K_{JC} is given in Kimura and Ohta (1972) in the following form:

$$V(K_{JC}) = \frac{p(1-p)}{L\left(1-\frac{4p}{3}\right)^2} \qquad (18.2)$$

where L is the length of the nucleotide sequences. Notice that all genetic distances here, except for Lake's paralinear distance, is expressed as the number of nucleotide substitutions per site, and is therefore independent of sequence length. The variance decreases with the increase in sequence length, as our intuition would have told us.

When nucleotide sequences have diverged so much that they reached full substitution saturation, then p is expected to be 0.75, and K_{JC} is infinite and cannot be represented in computer or in output. Whenever a genetic distance is too large to be coded by the computer, it is coded 9 and its variance is coded 0 in DAMBE. Note that K_{JC} is expressed as the number of substitutions per site, and is highly unlikely to reach a value as large as 9.

Jukes and Cantor's distance performs better in phylogenetic reconstruction than distances based on more complicated models when the nucleotide sequences are short (Zharkikh 1994). This is understandable because the JC model has the smallest number of parameters to estimate among all substitution models. The variance should consequently be smaller if the JC model is in fact the correct model of nucleotide substitution.

The JC69 distance assumes equal equilibrium frequencies of the four nucleotides, which is often not true. Tajima and Nei (1984) proposed a method that does not require this assumption. Their genetic distance and its large-sample variance are

$$K_{TN} = -b\ln\left(1-\frac{p}{b}\right) \qquad (18.3)$$

where

$$b = \frac{\left(1-\sum_{i=1}^{4} q_i^2\right)+\left(\frac{p^2}{h}\right)}{2} \qquad (18.4)$$

$$h = \sum_{i=1}^{3}\sum_{j=i+1}^{4}\frac{x_{ij}^2}{2q_iq_j} \qquad (18.5)$$

$$V(K_{TN}) = \frac{b^2 p(1-p)}{L(b-p)^2} \tag{18.6}$$

where x_{ij} is the proportion of pairs of nucleotides i and j between the two orthologous DNA sequences, and q_i is the equilibrium frequency of the ith nucleotide (i = 1, 2, 3, 4 corresponding to A, G, T, C).

Although K_{TN} is based on the equal input model in which the rate of substitution of a base to any other bases is assumed to be the same, regardless of the original nucleotide, the distance gives good estimates of the number of nucleotide substitutions for a variety of other substitution models (Nei 1987).

2.2 Kimura's two parameter distance

Kimura's (1980) distance is based on his two-parameter model, which was proposed to take into account the recognized fact that transitions and transversions occur at rather different rates, with the former occurring significantly more frequently than the latter. In protein-coding genes, this difference in substitution rate between transitional and transversional substitutions is caused mainly by transitions being mostly synonymous and transversions mostly nonsynonymous at the third codon positions (Xia et al. 1996).

Another cause of the transition bias is DNA methylation, which greatly elevates the C→T transition. Each C→T transition in one strand also leads to an A→G transition in the opposite strand, with the net effect of a much elevated rate of transitional mutations relative to that of transversional mutations.

The two-parameter model is symmetric and consequently necessitates an equal equilibrium frequencies of the four nucleotides. If we designate P as the proportion of transitional differences and Q the proportion of transversional differences between the two sequences, Kimura's distance and its sampling variance are expressed as:

$$K_{K80} = \frac{\ln(a)}{2} + \frac{\ln(b)}{4} \tag{18.7}$$

$$V(K_{80}) = \frac{a^2 P + c^2 Q - (aP + cQ)^2}{L} \tag{18.8}$$

where

$$a = \frac{1}{1-2P-Q}$$
$$b == \frac{1}{1-2Q} \tag{18.9}$$
$$c = \frac{a+b}{2}$$

With the increase of divergence time between the two sequences, P and Q would increase to approach 0.25 and 0.5, respectively. Because of stochastic effect, Q could occasionally be equal to, or even greater than, 0.5, and (2P + Q) could occasionally be equal to, or greater than, 1. Under such rare situations, K_{K80} is infinite and is coded as 9, and its variance coded as 0 in DAMBE.

2.3 F84 distance

The F84 distance is based on the nucleotide substitution model implemented in DNAML since 1984 (PHYLIP Version 2.6). It is based on the following substitution matrix with the four nucleotides arranged in the order of T, C, A, G:

$$Q = \begin{pmatrix} \cdot & (1+\frac{\kappa}{\pi_Y})\pi_C & \pi_A & \pi_G \\ (1+\frac{\kappa}{\pi_Y})\pi_T & \cdot & \pi_A & \pi_G \\ \pi_T & \pi_C & \cdot & (1+\frac{\kappa}{\pi_R})\pi_G \\ \pi_T & \pi_C & (1+\frac{\kappa}{\pi_R})\pi_A & \cdot \end{pmatrix} \tag{18.10}$$

where $\pi_Y = \pi_T + \pi_C$ and $\pi_R = \pi_A + \pi_G$. The element Q_{ij} $(i \neq j)$ represents the rate of substitution from nucleotide i to j. The diagonals Q_{ii} are specified by the mathematical requirement that row sums of **Q** are zero. Note that the model has three frequency parameters and one rate ratio parameter. Designate the observed transitions and transversions as P and Q, respectively, the F84 distance is calculated according to the following formula:

$$D_{F84} = 4\beta\,(\pi_T\pi_C * (1+\kappa/\pi_Y) + \pi_A\pi_G * (1+\kappa/\pi_R) + \pi_Y * \pi_R) \quad (18.11)$$

$$\beta = \frac{-\ln\left(1 - \dfrac{Q}{2\pi_Y\pi_R}\right)}{2} \quad (18.12)$$

$$\kappa = \frac{\alpha}{\beta} - 1 \quad (18.13)$$

$$\alpha = \frac{-\ln\left(\dfrac{2(\pi_T\pi_C + \pi_A\pi_G) + 2(\dfrac{\pi_T\pi_C\pi_R}{\pi_Y} + \dfrac{\pi_A\pi_G\pi_Y}{\pi_R})(1 - \dfrac{Q}{2\pi_Y\pi_R}) - P}{2\pi_T\pi_C/\pi_Y + 2\pi_A\pi_G/\pi_R}\right)}{2}$$

2.4 TN93 distance

The TN93 distance is based on the nucleotide substitution model proposed to accommodate heterogeneity between the two kinds of transitions, i.e., T↔C and A↔G (Tamura and Nei 1993). Empirical data suggest that T↔C transitions occur more frequently than A↔G transitions, which we have already encountered when counting pair-wise nucleotide substitutions in a previous chapter dealing with the pattern of nucleotide substitutions. These two rates of transitional substitutions are represented as κ_1 and κ_2 below:

$$Q = \begin{pmatrix} \cdot & \kappa_1\pi_C & \pi_A & \pi_G \\ \kappa_1\pi_T & \cdot & \pi_A & \pi_G \\ \pi_T & \pi_C & \cdot & \kappa_2\pi_G \\ \pi_T & \pi_C & \kappa_2\pi_A & \cdot \end{pmatrix} \quad (18.14)$$

Note that κ_1 and κ_2 are estimated as ratios of the transitional substitution rates over the transversional substitution rate, with the latter being set to 1. So the TN93 model has two rate ratio parameters. Designate the observed proportions of T↔C and A↔G transitions as P_1 and P_2, and the observed transversions as Q, then the TN93 distance is equal to

$$D_{TN93} = 4\pi_T\pi_C\kappa_1 + 4\pi_A\pi_G\kappa_2 + 4\pi_Y\pi_R\beta \tag{18.15}$$

$$\kappa_1 = \frac{-\ln\left(1 - \frac{\pi_Y P_1}{2\pi_T\pi_C} - \frac{Q}{2\pi_Y}\right) - \pi_R \ln\left(1 - \frac{Q}{2\pi_Y\pi_R}\right)}{2\pi_Y} \tag{18.16}$$

$$\kappa_2 = \frac{-\ln\left(1 - \frac{\pi_R P_2}{2\pi_A\pi_G} - \frac{Q}{2\pi_R}\right) - \pi_Y \ln\left(1 - \frac{Q}{2\pi_Y\pi_R}\right)}{2\pi_R} \tag{18.17}$$

$$\beta = \frac{-\ln\left(1 - \frac{Q}{2\pi_Y\pi_R}\right)}{2} \tag{18.18}$$

2.5 Lake's paralinear distance

Lake's (1994) paralinear distance is based on one of the most complicated model of nucleotide substitution. Let N_{ij} be the number of sites where the nucleotide is i in the first sequence and j in the second sequence and let J be the determinant defined by

$$J = \begin{vmatrix} N_{AA} & N_{AT} & N_{AC} & N_{AG} \\ N_{TA} & N_{TT} & N_{TC} & N_{TG} \\ N_{CA} & N_{CT} & N_{CC} & N_{CG} \\ N_{GA} & N_{GT} & N_{GC} & N_{GG} \end{vmatrix} \tag{18.19}$$

Lake's paralinear distance can now be expressed as

$$d_{Lake} = -\frac{1}{4}\ln\left(\frac{J}{\sqrt{F_1 F_2}}\right) \tag{18.20}$$

where

$$F_i = N_{A.i} N_{T.i} N_{C.i} N_{G.i} \tag{18.21}$$

with $N_{N.i}$ being the frequency of nucleotide N in sequence i (i = 1 and 2).

Lake's (1994) paralinear distance has three main advantages. First, it is additive. Second, it is general and can accommodate all heterogeneity in substitution rates among nucleotides. Third, it is presumably applicable in situations in which nucleotide frequencies change over time. The distance, however, also has two disadvantages. First, it is not explicitly expressed as the number of nucleotide substitutions per nucleotide site. Second, the distance is not always defined for two nucleotide sequences, and such inapplicable cases occur more frequently with Lake's distance than with other genetic distances implemented in DAMBE (Zharkikh 1994).

3. DISTANCES BASED ON CODON SEQUENCES

All distances from the previous section can also be computed for codon sequences because codon sequences are nucleotide sequences after all. However, all distances in the previous section are derived from nucleotide-based substitution models, which are inherently awkward in describing substitution patterns in protein-coding genes for the following reason. The substitution rate at nucleotide sites of a protein-coding gene depends not only on whether the substitution is a transition or transversion and whether the site is located in a functionally important segment or not, but also depends on codon-specific properties, such as which codon position the site is at, whether the substitution is synonymous or nonsynonymous, and how similar the two coded amino acids are to each other when the substitution is nonsynonymous. (Xia 1998b; Yang 1996b). In short, a more realistic codon-based model is needed to handle the complexity of substitutions involving codons.

A good substitution model should incorporate two categories of variables that affect the rate of substitution. The first category is referred to as frequency parameters. For example, if nucleotide sequences are extremely GC rich, then almost all substitutions we observe will be G↔C transversions. For codon substitutions, we need consider not only nucleotide frequencies, but also codon frequencies. Take the universal genetic code for example, in the extreme case when an ancestral protein is made of entirely of methionine (coded by AUG only), then any nucleotide substitution will lead to a nonsynonymous substitution. If we ignore codon frequencies, we may conclude, based on the observation that nonsynonymous substitutions far outweigh synonymous substitutions, that the sequences are under strong positive selection. Such conclusions have in fact been made numerous times in literature without any reference to codon frequencies.

The second category of parameters are rate parameters or rate ratio parameters (because we can estimate only the relative rate but not the absolute rate). The most important factor determining the rate parameter of codon substitution is how much the amino acid coded by the new codon differ from the amino acid coded by the original codon. If the substitution is synonymous, then the difference is zero. If the substitution is nonsynonymous, then the substitution rate depends much on the similarity in physico-chemical properties of the two amino acids involved.

There are two categories of methods that calculate synonymous and nonsynonymous substitution rates while taking into consideration amino acid dissimilarities. The first is what I call the empirical counting approach, and the second is the model-based maximum likelihood approach.

3.1 The empirical counting approach

Five methods (Li 1993; Li et al. 1985b; Miyata and Yasunaga 1980; Nei and Gojobori 1986; Perler et al. 1980) belong to this category. In general, they all involve counting the number of synonymous and nonsynonymous sites, and synonymous and nonsynonymous nucleotide substitutions from codon by codon comparisons. The methods differ in two perhaps minor aspects. First, some methods (Nei and Gojobori 1986; Perler et al. 1980) give an equal weight to different codon substitution pathways. Let me explain what a pathway is. When we have two codons differing at a single position, e.g., AAA and AAG, then there is only one simplest pathway for one codon to change into the other. When two codons differ by more than one position, e.g., TTT and GTA, then we have two simplest pathways (Nei 1987, p.75):

Pathway I: TTT (Phe) ↔ GTT (Val) ↔ GTA (Val)
Pathway II: TTT (Phe) ↔ TTA (Leu) ↔ GTA (Val)

Pathway I involves one synonymous substitution and one nonsynonymous substitution, whereas Pathway II involves two nonsynonymous substitutions. The question is whether we should consider these two pathways equally likely or not. Methods that treat different pathways as equally likely (Nei and Gojobori 1986; Perler et al. 1980) are called unweighted pathway methods, otherwise they are called weighted pathway methods (Li 1993; Li et al. 1985b; Miyata and Yasunaga 1980).

As we know that nonsynonymous substitutions are much less likely to happen than synonymous substitutions, weighted pathway methods seem to be more reasonable. However, the difference is the resulting estimates is rather small because in many cases alternative pathways are similar. Take codons AAA and ACG for example. We have two simplest pathways:

Pathway I: AAA (Lys) ACA (Thr) ACG (Thr)
Pathway II: AAA (Lys) AAG (Lys) ACG (Thr)

Both pathways involve one synonymous substitution and one nonsynonymous substitution between lysine and threonine. It is therefore quite reasonable to give an equal weight to the two pathways.

Another difference between the methods is the correction for multiple hits. Some (e.g., Nei and Gojobori 1986) uses the Jukes-Cantor one-parameter model to correct for multiple hits, whereas others (e.g., Li 1993) uses Kimura's two-parameter model. The method by Li et al. (1993) is implemented in DAMBE.

One should be aware that Kimura's two-parameter model cannot be used in all situations for correcting for multiple hits. When Kimura's two parameter correction is not applicable, the JC69 correction is applied in DAMBE. There are two situations when Kimura's two-parameter correction is not applicable: (1) when $Q = 0.5$, in which case there is no justification (no need) for Kimura's two-parameter method, and (2) when $(2P + Q)$ is equal to, or greater than, 1. When substitution saturation is so severe that p approaches 0.75, then even JC69 correction is inapplicable. DAMBE will inform you of the problem, and the incalculable distance will be set to be 1.2 times the largest calculable distance in the matrix.

As I mentioned before, a good codon-based method should consider both the frequency parameters and the rate ratio parameters. All the empirical counting methods that I have mentioned ignore the frequency parameters, i.e., they do not adjust for biased nucleotide frequencies or codon frequencies. If you recall, both the JC69 model and the K80 model used in these empirical counting methods do not take into account frequency parameters, and the TN84 model (Tajima and Nei 1984) is the simplest model that incorporates frequency parameters (nucleotide frequencies but not codon frequencies). In previous chapters, we have learned that nucleotide usage and codon usage are both typically biased. The unequal nucleotide and codon frequencies may well bias the estimate of synonymous and nonsynonymous substitutions. Some progress has been made to remedy this problem (Ina 1995; Moriyama and Powell 1997).

3.2 Codon-based maximum likelihood method

Two codon-based models (Goldman and Yang 1994; Muse and Gaut 1994) have been proposed and implemented in phylogenetic analysis by maximum likelihood. They can potentially be used for estimating synonymous and nonsynonymous substitution rates. One model (Muse and Gaut 1994) is restrictive in two ways. First, it does not have separate rate parameters for transitions and transversions. Second, it assumes that nonsynonymous substitutions occur equally likely. For example, two very different nonsynonymous substitutions, one involving AAT↔GAT (resulting in Asn↔Asp, with a Grantham's distance = 23 between the two amino acids) and the other involving TGT↔CGT (resulting in Cys↔Arg, respectively, with Grantham's distance = 180), were considered by the model to have the same substitution rate.

This assumption is known to be false. Amino acid substitutions occur more frequently between similar amino acids than between dissimilar ones (Clarke 1970; Epstein 1967; Grantham 1974; Kimura 1983, p. 152; Miyata et al. 1979; Sneath 1966; Xia and Li 1998; Zuckerkandl and Pauling 1965). This has been verified empirically in previous chapters dealing with codon and amino acid substitution. A codon substitution model that ignores this relationship is not a realistic model.

The other codon-based model (Goldman and Yang 1994), commonly referred to as the GY94 model, accommodates potentially different substitution rates among different amino acid replacements by using dissimilarity measures between amino acids. This model does not fit empirical codon substitution data well, and has subsequently been modified into a more general geometric relationship (Yang et al. 1998) which I will referred to as the YRH98 model.

The YRH98 model uses dissimilarity indices between amino acids to accommodate the rate heterogeneity among different kinds of amino acid substitutions. In this aspect it is the same as the GY94 model. There are currently two amino acid dissimilarity indices in use, Grantham's distance (Grantham 1974) and Miyata's distance (Miyata et al. 1979). The GY94 model used Grantham's distance, whereas Miyata's distance was claimed to fit codon substitution data better than Grantham's distance for mammalian mitochondrial DNA (Yang et al. 1998). The YRH98 model consequently uses Miyata's distance as a measure of amino acid dissimilarity.

There is a major problem with the existing codon-based model (including the most recent YRH 98 model). Suppose we have a number of arginine and glycine codons (coded by CGN and GGN codons, respectively) in the ancestral sequence. Also suppose that the protein domain harbouring these arginine and glycine residues happens to be unimportant for normal function of the protein. Consequently, CGN and GGN codons would substitute each other at a nearly neutral rate. Because arginine and glycine are rather different amino acids, all codon-based models incorporating amino acid dissimilarities would predict that arginine codons and glycine codons should rarely mutate into each other. Consequently, these models will greatly overestimate the nonsynonymous substitution rate for our fictitious but not unrealistic example. Note that this problem is shared by the empirical counting methods presented in the previous section that adjusted the estimates by amino acid dissimilarities. DAMBE offers a graphic display of substitutions over sites for detecting conserved or variable domains.

Codon-based models can easily incorporate both the frequency and rate parameters. However, because of the large number of parameters, a codon-based maximum likelihood method is necessarily slow, and will most likely suffer from the undesirable consequence of overfitting the model if the sequences are not extremely long. This is a major problem because a typical protein-coding gene has just a few hundred codons. If one concatenate different protein-coding genes together to increase sample size, then one runs into the problem of obscuring the heterogeneity among the genes. For example, if some genes are from GC-rich isochores and others from AT-rich isochores, then the concatenation of these genes together will simply obscure the distinct evolutionary dynamics of these two categories of genes and defeat the original purpose of incorporating the frequency parameters. Similarly, if some genes are important and evolve slowly while others are unimportant and evolve fast, the concatenation of these genes defeats the original purpose of incorporating different rate parameters.

4. DISTANCES BASED ON AMINO ACID SEQUENCES

There are two measures of genetic distance based on amino acid sequences. The first is based on the Dayhoff PAM matrix, and the second is Kimura's (1983) empirical distance expressed in the following form

$$D = -\ln(1 - p - 0.2p^2) \qquad (18.22)$$

where p is the fraction of amino acids differing between the two sequences.

Both distances are not difficult to compute, but I have strong reservation about their use. There are great rate differences among different amino acid substitutions. Amino acid substitutions occur more frequently between similar amino acids than between dissimilar ones (Clarke 1970; Epstein

1967; Grantham 1974; Miyata et al. 1979; Sneath 1966; Xia and Li 1998; Zuckerkandl and Pauling 1965). Kimura's formulation does not take this rate heterogeneity into consideration, and the equation is derived from a biased sample of substitution data.

Various approaches have been proposed to accommodate the rate heterogeneity among different amino acid substitutions, such as the use of empirical transition models for amino acids (Dayhoff et al. 1978) or use a mechanical substitution model based on transition probabilities derived from empirical substitution data (Adachi and Hasegawa 1996). Both assume that the empirical transition model for amino acids is sufficiently general to be applied to all proteins. Given the large variation in substitution rates among genes (Wu and Li 1985) and among lineages (Gaut et al. 1992), this assumption is almost certainly false.

Different amino acids can differ in many physico-chemical properties; indeed, 134 properties were enumerated (Sneath 1966). Xia and Li (1998) studies 10 amino acid properties, and all have significant effect on substitution rate of codons, and some appeared to have played a significant role in the evolution of genetic code and amino acid composition. A good estimate of genetic distance based on amino acid sequences need to take into account the differences in physico-chemical properties between amino acids.

5. GENETIC DISTANCES FROM ALLELE FREQUENCIES

This section deals with estimating genetic distances by using allele frequency data. Three commonly used genetic distances can be computed by using DAMBE: Nei's (1972) distance, Cavalli-Sforza's (1967) chord measure, designated as D_{CE}, and Reynolds, Weir, and Cockerham's (1983) genetic distance, designated as D_{RWC}. All three assume that genetic divergence between OTUs arise from genetic drift, and assumptions specific to each distance measure are listed in sections dealing with individual distances.

The three genetic distances are based on models of genetic drift and were intended for measuring genetic divergence between populations. The use of these genetic distances for measuring genetic divergence between species requires another major assumption (in fact a major jump of faith) that the genetic variation we observe within populations, between population or between species are really of the same variation, shaped by the same evolutionary process. This assumption gained increasing acceptance with the increased popularity of the neutral theory of molecular evolution (Kimura 1983). However, the assumption may not be true. For example, if the genetic variation within and between populations is mainly shaped by genetic drift, whereas speciation is caused by rapid fixation of one or very few highly favourable mutations, then the three genetic distances clearly would not be appropriate for measuring divergence between species.

One disadvantage shared by all these distances is the use of allele frequencies rather than genotypic frequencies, which contains much more information. For example, suppose we have a locus with two alleles, with allele frequencies being p_1 and p_2, respectively. The information contained in the allele frequencies is

$$H_A = -\sum_{i=1}^{n} p_i \log_2 p_i = -[p_1 \log_2 p_1 + p_2 \log_2 p_2] \qquad (18.23)$$

where $\log_2$ stands for logarithm with base 2, and the unit for H_A is bit in information science, with eight bits making a byte. The information in the genotypic frequencies, assuming Hardy-Weinberg equilibrium, is

$$H_G = -[p_1^2 \log_2 p_1^2 + 2p_1p_2 \log_2 (2p_1p_2) + p_2^2 \log_2 p_2^2] \qquad (18.24)$$

When p_1 approaches 0 or 1, then both H_A and H_G will approach 0. When p_1 is between 0 and 1, then H_A is always smaller then H_G. For example, if $p_1 = 0.5$, then $H_A = 1$ and $H_G = 1.5$. The loss of information could therefore be very substantial.

Students often question the genetic distances by the following example. Suppose one population has genotypic frequencies for the three genotypes, AA, Aa and aa, equal to 1000, 0, and 1000, whereas the corresponding values from another population is 0, 1000, 0. Both would then have the same allele frequencies which completely obscure the difference between the two populations. In response to this critique, we note that the first population is under strong selection against heterozygotes and the second population is under strong selection against homozygotes. These kinds of selection are assumed to be either absent or so rare as to be negligible in evolutionary models underlying the genetic distances.

5.1 Nei's genetic distance:

Nei's genetic distance (Nei 1972) is formulated in the following form:

$$D_{Nei} = -\ln\left(\frac{\sum_{i=1}^{m}\sum_{j=1}^{n_i} p_{ij}q_{ij}}{\sqrt{\left(\sum_{i=1}^{m}\sum_{j=1}^{n_i} p_{ij}^{2}\right)\left(\sum_{i=1}^{m}\sum_{j=1}^{n_i} q_{ij}^{2}\right)}}\right) \tag{18.25}$$

where n_i is the number of alleles for locus i and m is the number of alleles surveyed, p_{ij} is the frequency of allele j for locus i for OTU 1 and q_{ij} is the corresponding frequency for OTU2. D_{Nei} can take values from 0 to infinite. Because we cannot take logarithm of values equal or smaller than 0, the numerator cannot be zero. This implies that the two OTU's cannot be completely different. For example, suppose we surveyed 5 loci, each with two alleles. If OTU 1 is fixed for one allele in all 5 loci, whereas OTU 2 is fixed for an alternative allele for all 5 loci, then Nei's distance is infinite and cannot be represented in the computer. In practise, such situations must be extremely rare unless one sampled very few loci.

Nei's distance is formulated for an infinite-isoallele model of mutation, in which there is a rate of neutral mutation and each mutant represents a new allele. For protein electrophoretic data, these assumptions may not be true. For example, protein mobility depends on size, shape and charge of the protein molecule. If one mutation increases the charge, another mutation may decrease the charge. The mobility of the new protein is then reversed back to the original state, with the consequence that the new mutant is not scored as a new allele. Ohta and Kimura (1973) proposed a stepwise-mutation model which might be more appropriate for electrophoresis data.

Nei's distance also has other assumptions that are difficult to assess, such as equal rates of neutral mutation for all loci, equilibrium between mutation and genetic drift for the genetic variability initially in the population, and constant population size.

Nei et al. (1983) proposed another distance that seems to perform well in phylogenetic analysis. It is expressed as

$$D_{Nei83} = \frac{\sum_{i=1}^{m}\left(1-\sum_{j=1}^{n_i}\sqrt{p_{ij}q_{ij}}\right)}{m} \tag{18.26}$$

This distance takes the value between 0 and 1 and therefore is not linearly related to the number of gene substitutions. However, when D_{nei83} is small, it increases roughly linearly with evolutionary time. This distance is not implemented in DAMBE.

5.2 Cavalli-Sforza's chord measure

The cord measure (Cavalli-Sforza and Edwards 1967), designated as D_{CE} is expressed as

$$D_{CE} = \frac{4\sum_{i=1}^{m}\left(1-\sum_{j=1}^{n_i}\sqrt{p_{ij}q_{ij}}\right)}{\sum_{i=1}^{m}(n_i-1)} \tag{18.27}$$

D_{CE} assumes that there is no mutation, and that all gene frequency changes are by genetic drift alone. However, it does not assume that population sizes have remained constant and equal in all populations, as does Nei's distance. It copes with changing population size by having expectations that rise linearly not with time, but with the sum over time of 1/N, where N is the effective population size. This seems to be a rather awkward ways of relaxing the assumption of the constant populations size, because, by so doing, the distance will no longer be linearly related to evolutionary time (Latter 1972).

5.3 Reynolds, Weir, and Cockerham's genetic distance

Reynolds, Weir, and Cockerham's (1983) genetic distance, designated as D_{RWC} is expressed as

$$D_{RWC} = \frac{\sum_{i=1}^{m}\sum_{j=1}^{n_i}(p_{ij}-q_{ij})^2}{2\sum_{i=1}^{m}\left(1-\sum_{j=1}^{n_i}p_{ij}q_{ij}\right)} \tag{18.28}$$

Because it is just a different estimator of the same parameter as D_{CE}, the two distances share similar assumptions. Nei (1987), Li (1997), and the documentation for PHYLIP (Felsenstein 1993) provide illuminating discussions of various genetic distances based on allele frequencies. This chapter is in a large part a summary of their discussions on genetic distances.

One should be aware of the fact that, when these distances are to be inputted into computer programs for phylogenetic reconstruction, these phylogenetic programs have additional assumptions about these distances. For example, the statistical model underlying the Fitch-Margoliash method and neighbour-joining method implicitly assumes that the distances in the input distance matrix have independent errors. This obviously cannot be true because OTU's are phylogenetically related and share some ancestral lineages for various periods.

Chapter 19

Molecular Phylogenetics: Concepts and Practice

Molecular phylogenetics has been developed to solve two related problems. The first is to infer the branching pattern of different species from their common ancestor, and the second is to infer when such branching events, or cladogenic speciation events, have occurred during geological time. Three categories of methods are commonly used for these two purposes: distance methods, maximum parsimony methods, and the maximum likelihood methods. These methods have been implemented in DAMBE. This chapter covers the fundamentals of using these methods.

The DNA in an organism is like a very long book, which has been passed on along the germ line to the present generation from time immemorial. Whenever a cell replicates, the book is copied from the beginning to the end. Some copying errors (mutations) would occur during the copying process. If such a mutation becomes fixed in the species, then the species has diverged from its closest relative by one substitution. It is these changes recorded on the DNA book that we can use to infer phylogenetic relationships among organisms sharing a common ancestor.

Molecular phylogenetics has been developed rapidly, and its application has already generated many successful stories. One of the successes that may come to your mind is perhaps the establishment of the three-kingdom classification. However, there is a much more interesting success story that you may not have heard of. It is this story that I will tell you now.

Off the western coast of South America there is a volcano island called Chiloé Island on which a special kind of fox, named Darwin's fox (*Dusicyon fulvipes*), was found. On the mainland opposite the island thrives another fox species, the gray fox (*Urocyon cinereoargenteus*). For a long time it has been thought that Darwin's fox is descended from the gray fox. In other words, when the volcano island was formed, some gray foxes have somehow got onto the island. They then diverged independently to become Darwin's fox after the seawater rose isolating the island from the mainland.

In 1980s when molecular techniques became widely available to field biologists, researchers began to reconstruct phylogenetic trees for various fox species and dating their speciation events. They were surprised to find that the divergence time between Darwin's fox and the gray fox were much earlier than the formation of the volcano island. How could this happen? If Darwin's fox is descended from the gray fox, after the formation of the island, then the divergence time between the gray fox and Darwin's fox should not predate the formation of the island.

One possible hypothesis is that Darwin's fox had diverged from the gray fox a long time ago on the mainland, long before the island was formed. After the formation of Chiloé Island, some Darwin's foxes, not gray foxes, migrated to the island and became established. Meanwhile, the mainland population had gone extinct.

This is a bold hypothesis. It predicted the existence of a species on the mainland that nobody had seen. However, researchers had faith in the prediction and went on looking for historical footprints (e.g., fossils) of Darwin's fox left on the mainland. It is in search of these footprints that researchers found themselves face to face with a living population of Darwin's fox on the mainland. What a reward! They have boldly predicted the existence of Darwin's fox on the mainland, at least in the past. Now the prediction is confirmed with a pleasant surprise. This discovery, in my opinion, rivals the success of predicting the existence of an unseen planet based on the orbits of other visible planets.

Although the real story is not exactly like this, all the essential elements are there. May you make the same or even more exciting discovery!

1. THE MOLECULAR CLOCK AND ITS CALIBRATION

A molecular clock is in some way similar to a conventional mechanical clock. We measure the lapse of time with a mechanical clock by counting the number of ticks. With a molecular clock, we measure the time by counting the number of substitutions of nucleotides or amino acids, which is measured by various kinds of genetic distances detailed in a previous chapter. Each substitution is equivalent to one tick in a mechanical clock. The more substitutions, the longer the elapsed time is.

A mechanical clock is read with reference to midnight, which is defined to be zero o'clock. When it is 12 o'clock, we know that 12 hours have elapsed since midnight. With a molecular clock we have no idea about its initial state. So the state of the molecular sequence by itself does not allow us to read time from it. For this reason we assume that all living beings on earth share a common ancestor and all molecular clocks in different organisms begin to tick in their own way since the divergence of organisms from their common ancestor (fig. 1). In this way, we can compare the molecular sequences from different organisms and count the number of sites differing between organisms. This number of different sites divided by 2 is the crudest estimate of the number of ticks in a molecular clock.

AAACCCCGGGGCCCCTATTTTTTG

AA**G**CCCCGGGGCCCCTATTTTTTG	AAACCCCGGGGCCCCTATTTTTT**T**
AAGCC**T**CGGGGCCCCTATTTTTTG	AA**T**C**T**CCGGGGCCCCTATTTTTTT
AAGCCTCGGGGCCC**T**TATTTTTTG	AATCTCCGGGGCC**T**CTATTTTTTT

Figure 1. Sequence evolution. The top sequence is the ancestral state, which is not visible to us, and the bottom two sequences are the current states of the two descending lineages. The line in the middle indicates independent evolution of the two lineages. Nucleotide substitutions have been bolded.

Let's look at the sequence evolution depicted in Fig. 1, The sequence is of 24 bases long. The information available to us are the two bottom sequences, representing the current states of the two diverging lineages. The comparison of the two bottom sequences site by site yields seven sites that are different between the two sequences. So on average, the two sequences have diverged for 3.5 (=7/2) nucleotide substitutions along each genealogical lineage. If expressed in the number of substitutions per site, then we have a genetic distance between the two species being 7/24 substitution per site.

Can you use the two sequences on the bottom of fig. 1 to reconstruct the ancestral sequences on the top, with the criterion that the maximum number of substitutions being no more than seven? Do you find just a single sequence, or several sequences, satisfying the criterion?

1.1 Calibrating a molecular clock

A statement that two species have diverged for 3.5 nucleotide substitutions is not as satisfying as a statement that they have diverged for two million years. How to translate these 3.5 nucleotide substitutions into years or millions of years? This translation is achieved by calibrating the molecular clock.

Before calibrating a molecular clock, let's have a short digression into geology. It is generally true that sedimentary rocks form on top of older rocks, often with fossils buried inside. If fossils of rats and mice are found in one stratum, but not in any older strata everywhere on earth, then we may assume that it is during the period when that sedimentary stratum was formed when ancestors of mice and rats diverged. If the stratum is found to be one million years old, we can infer that mice and rats must have diverged one million years ago. This allows us to fill in a value of 1 in the first cell of the second column in Table 1. The same kind of geological dating would allow us to fill in a number of cells in the second column in Table 1. There is one pair of species, numbered 10, that do not have any fossil record. How many million years have these two taxa diverged from each other?

Table 1. Data for calibrating a molecular clock. The divergence time is in million years. T_D is divergence time from geological dating, and D_G is genetic distance from molecular data.

Species Pair	T_D	D_G
1	1	0.05
2	2	0.12
3	3	0.11
4	4	0.2
5	5	0.21
6	6	0.27
7	7	0.35
8	8	0.33
9	9	0.41
10	?	0.29

The data in Table 1 would allow us to calibrate our molecular clock for estimating the divergence time between the two species in the 10th pair. Now for each species pair, we can obtain the DNA sequences from orthologous genes and compute genetic distances, which are shown in the last column of Table 1.

Plotting the genetic distances versus the divergence time obtained from geological dating leads to fig. 2. We are pretty lucky that the relationship between T_D and D_G is roughly linear. A simple linear regression, with the intercept forced to be zero, yields

$$T_D = 21.98 D_G \tag{19.1}$$

where T_D is the divergence time and D_G is the genetic distance. The equation means that a genetic distance of 1 is equivalent to 21.98 million years. For our species pair #10, the divergence time equals 6.37 (=21.98 * 0.29) million years. In summary, as long as we have a number of species pairs with well-dated fossil record, we can obtain divergence time from a molecular clock for any given pair of species that left no trace in the fossil record.

There are some complications involving dating speciation event, some being related to genetic distances and some to the calibration procedure. The calculation of various genetic distances have been detailed in a previous chapter. Here we briefly outline problems associated with the calibration.

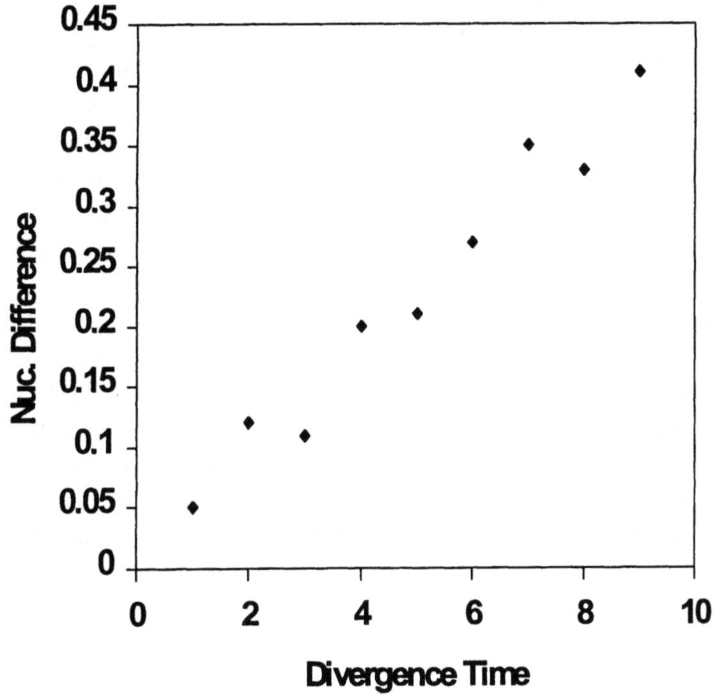

Figure 2. Calibration of a molecular clock.

1.2 Complications in calibrating a molecular clock

There are three main complications in calibrating molecular clock. The first is the generation time, which manifests at the genome level. The second is the differential cell-replication rate between male and female germ lines, which manifests at chromosome level, and will be referred to as germ-line effect. The third is the effect of genetic context, which manifests at the gene level.

1.2.1 The generation time effect

It is important to note that most mutations occur when the DNA is duplicated, especially during meiosis. Some lineages have short generation time, e.g., *E. coli* replicate once every 20 minutes in favorable conditions, while some others, such as human, have relatively long generations. The DNA in the former is consequently duplicated more frequently, and therefore is expected to experience many more changes, than the latter given the same period of time. Consequently, the calculated genetic distance, which is typically expressed as the number of substitutions per site for molecular sequences, will also be much greater in the former than in the latter given the same evolutionary time. Thus, it is possible that a genetic distance of 0.2 may be equivalent to one million years in the *E. coli* lineage but be equivalent to three million years in the lineage leading to human. If the nine species pairs with the known geological time are all mammalian species, whereas the 10th species pair are two bacterial species, then the divergence time between the two bacterial species may be severely overestimated.

The effect of the generation time effect has been demonstrated by comparing evolutionary lineages with different generation time, e.g., between humans and monkeys (Ellsworth et al. 1993; Seino et al. 1992), and between rodents and primates (Gu and Li 1992). For this reason, whenever we are interested in dating the speciation events between two or more species (i.e., target species), we should always calibrate the molecular clock by using data from species pairs that are phylogenetically similar to the target species.

1.2.2 The germ-line effect

In mammalian species, mutation rate is much higher in the male germ line than in the female germ line because the male germ line experiences many more cell replications than the female germ line. Human germ cells during spermatogenesis divide continuously and reach a number of approximately $1.2 * 10^9$ Ad spermatogonia at about 13 years of age (Vogel and Motulsky 1986). This requires about 30 cell generations of dichotomous divisions with no cell loss. Ad spermatogonia undergo continuous divisions and divide once about every 16 days. One of the two products of the division continues to produce Ad spermatogonia whereas the other gives rise to two Ap spermatogonia that, after 3 more mitotic divisions and 2 divisions involved in meiosis, become haploid spermatids. This implies that the number of cell generations (designated by n) a sperm has left behind depends on the age of the male and follows the following equation:

$$n \approx 30 + (Age - 13) \cdot (365/16) + 6 \approx -260.6 + 22.8Age \qquad (19.2)$$

for Age > 13.

Because n increases with the age in men, we expect that the number of mutations accumulated in the male germ cells should also increase linearly with increasing age. The prediction is supported by empirical data on human genetic diseases. An extensive review (Vogel and Motulsky 1986, sections 5.1.3.3) revealed that (1) the mutation load accumulated in male germ cells increased with age of the male and (2) the mutation load increased with the age of the male at an increasing rate.

In contrast to the male germ line, the germ line of the human female experiences cell proliferation only in the early part of her life, with the number of cells increasing to a peak of about 7 millions in the fifth month. Few cells are eliminated during this proliferation process. This requires approximately 23 dichotomous cell divisions, with no cell loss.

Equation (19.2) shows that the number of cell divisions a sperm has undergone from early embryonic development up to the age of 28 is about 378, which is about 16 times greater than the number of divisions in the life history on an egg. Consequently, such a sperm would accumulate many more mutations as the egg would have accumulated. Molecular evolution in mammals is indeed mostly male-driven, with males serving as a major source of nearly neutral mutations (Miyata et al. 1987). Interestingly, for conserved (presumably functionally important) genes, the amount of mutations contributed by the male is not different from that contributed by the female. This indicates that selection is more stringent on the male germ line than on the female germ line.

The reasoning above suggests that the genome in an average sperm in human male is poorer than that in an egg, which allows us to understand why the environment of the reproductive tract of females is very unfriendly to the sperm. Take human being for example. After entering the reproductive tract of the female, the sperm has to survive the acidic environment in the vagina of the female, swim vigorously against the downward flow in the female reproductive tract, and escape the attack of the self-defence system of the female. Half of the sperm is killed instantly upon entering the female reproductive tract. Only about 50 out of 200 million sperms may eventually reach the egg and only one can secure the fertilization. We can interpret those barriers in female reproductive tract as selection mechanisms to filter out sperms of poor genotypes. This also corroborates the finding that selection is more stringent against male gametes than against the female gametes.

What does this have to do with the calibration of the molecular clock? We expect the genes on the Y-chromosome to evolve at a much faster rate than those on the autosomes. If one gene that was originally located on the Y-chromosome has relocated to other chromosome in one lineage, but remained on the Y-chromosome in other lineages, then we should expect the gene to tick much faster when carried on the Y-chromosome than its counterpart relocated to other chromosomes.

1.2.3 The Effect of Genetic Context

Vertebrate genomes have GC-rich and AT-rich isochores. Mutation spectrum is GC-biased in the former but AT-biased in the latter. If one gene originally located in an AT-rich isochore subsequently moved to a GC-rich isochore, then it will experience many AT to GC mutations. Such changes in genetic context will result in very different evolution rates among different lineages. We have previously encountered a case in which the DNA sequences of the elongation factor-1α gene differ greatly in nucleotide frequencies. It is likely that these DNA sequences have evolved in different genetic context and one should be cautious when using them to make phylogenetic inferences.

It would seem appropriate to show you how to test the molecular clock hypothesis at this point. However, the test requires the knowledge of phylogenetic relationships among the lineages. Even for the simplest relative rate test, you still need to specify which taxon should be the outgroup. Although it is very simple to do the test in DAMBE, I prefer to delay its introduction to later chapters, after you have learned how to use DAMBE for phylogenetic reconstruction.

2. COMMON APPROACHES IN MOLECULAR PHYLOGENETICS

There are many approaches to molecular phylogenetic reconstruction, but most can be classified into one of three categories: the distance methods, the maximum parsimony methods and the maximum likelihood methods. In this section you will learn how to use DAMBE to reconstruct phylogenetic trees using any one of these methods. Testing the molecular hypothesis, as well as statistical evaluation of alternative phylogenetic hypotheses, will be dealt with in later chapters.

2.1 Distance methods

Distance methods operate on a genetic distance matrix, which are typically calculated on the basis of certain substitution model. Because nucleotide sequences and amino acid sequences differ in substitution patterns, different genetic distances have been proposed for them. Within nucleotide sequences, some code for proteins and some do not (e.g., introns, rRNA, etc.). The former is often referred to as codon sequences and differ from the latter in substitution patterns. They consequently are associated with different genetic distances.

Commonly used genetic distances for non-protein-coding nucleotide sequences are Jukes and Cantor's (1969) one-parameter distance, Kimura's (Kimura 1980) two-parameter distance, Tajima and Nei's (1984) distance, the F84 distance (Felsenstein 1993), the Tamura and Nei's (Tamura and Nei 1993) distance, and Lake's (1994) paralinear distance. Read the chapter on genetic distances for further information on these genetic distances. These distances are all based on nucleotide-based models.

It is known that substitution rates vary much over sites, and such rate heterogeneity would result in underestimation of genetic distances. A gamma distribution has often been used to correct for such underestimation. The gamma-corrected distances are implemented in DAMBE for the F84 and the TN93 models.

The genetic distances above can also be computed for codon sequences. However, nucleotide-based models and distances from such models are inherently awkward in describing substitution patterns in codon sequences because the substitution rate at nucleotide sites of a protein-coding gene depends not only on whether the substitution is a transition or transversion and whether the site is located in a functionally important segment or not, but also depends on codon-specific properties, such as which codon position the site is at, whether the substitution is synonymous or nonsynonymous, and how similar the two amino acids are to each other when the substitution is nonsynonymous. (Xia 1998b; Yang 1996b). In short, the nucleotide-based substitution model cannot handle the complexity of substitutions involving codons and codon-based substitution models are called for.

Various genetic distances, including those based on nucleotide, codon, and amino acid sequences, as well as those based on allele frequencies, have been reviewed in the chapter dealing specifically with genetic distances. Please refer to that chapter for further information.

Sometimes it is argued that distance methods can use more information in the sequences. For example, if we have six OTUs with one of them having a deletion of 10 bases from sites 20-30, then maximum likelihood methods, being site-oriented, will simply delete sites 20-30 for all sequences. Thus we lose all information contained in the stretch of 10 sites. With the distance methods, the pair-wise distance can still be computed between any two of the five OTUs with no deletions. However, there is one major problem with this approach. If that stretch of 10 bases happens to be highly variable, and all the rest of the sequences high conservative, then we will have a small genetic distance between the sequence with the deletion and all other sequences without the deletion, but a large genetic distance between any two from the five sequences without the deletion. This difference in genetic distances is an artefact that will bias phylogenetic estimation. It is always a good idea to do a site-wise deletion of all unresolved sites and gaps.

2.1.1 DAMBE and distance methods based on molecular sequences

Start DAMBE, and open a nucleotide sequence file, such as the **invert.fas** file that comes with DAMBE. Click **Phylogenetics|Distance methods|Nucleotide sequences**. A dialog box appears (fig. 3) for you to select options.

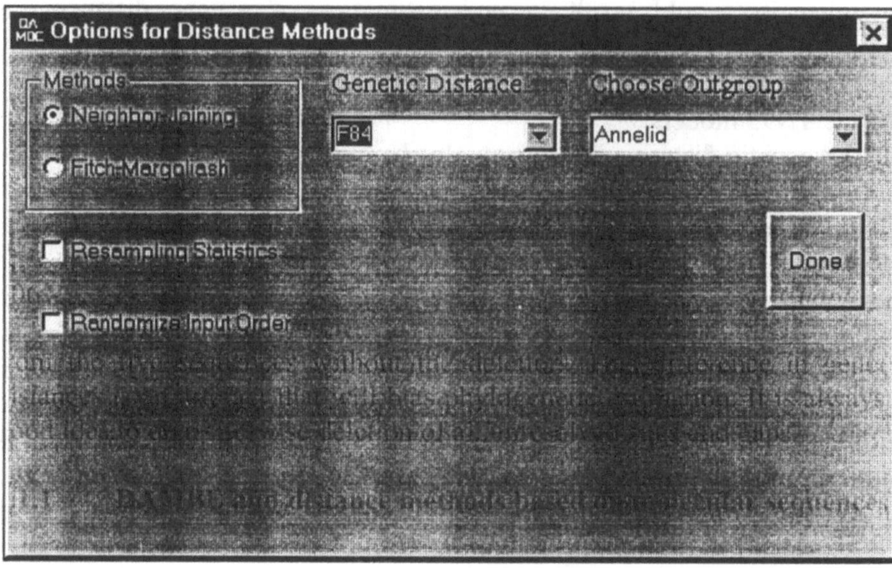

Figure 3. Dialog box for selecting options in phylogenetic analysis using distance methods. More options and checkbox will appear when click different option buttons or checkboxes.

You may wonder why the bottom part of the dialog box is blank. It is not. Relevant options and checkboxes will appear in response to your choices. I will explain all the option buttons, checkboxes, and dropdown menus in a moment. You may click the **Choose outgroup** dropdown menu to select **annelid** as an outgroup, and click the **Genetic distance** dropdown menu to select **F84**, which is the nucleotide-based substitution model implemented in PHYLIP' DNAML program since 1984. An additional input field, labelled **Alpha value**, will appear when you choose the F84 or TN93 distances. This is for you to input an alpha value to correct the underestimation of genetic distances when there is substantial rate heterogeneity over sites. Alpha is called the shape parameter in a gamma distribution. Its default value is 0, which actually means infinite. The zero is used simply because a computer cannot represent an infinitely large value. An infinitely large alpha value means that there is no rate heterogeneity, and a decreasing alpha value corresponds to increasing rate heterogeneity. Because we have not yet learned how to estimate the alpha parameter, just leave the field as is and click the **Done** button. A phylogenetic tree will be displayed in the tree window (fig. 4)

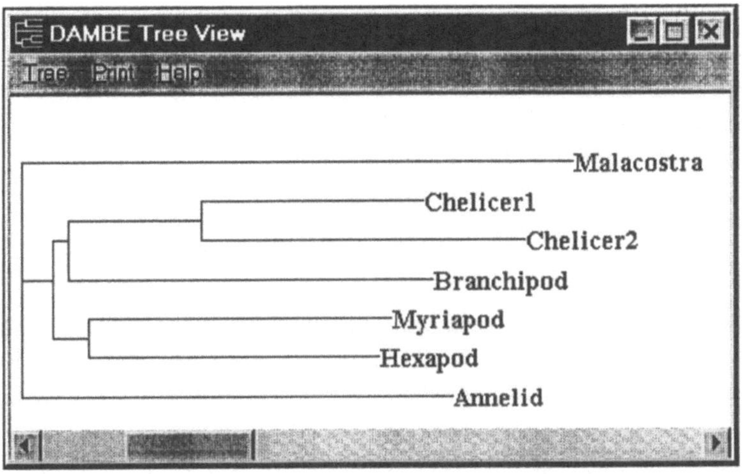

Figure 4. Display window for tree manipulation and printing.

Some basic tree manipulations, such as tree re-rooting, node numbering, font and line thickness changes, etc., are implemented. The tree can also be printed in high quality to a printer attached to your computer.

Other information related to tree reconstruction, such as what genetic distance has been used, what options you have chosen, which species is used as outgroup species, and the matrix of genetic distances used in reconstruction, is shown in DAMBE's display window. Also shown in the display window is the tree written in PHYLIP format. A sample of the text display is shown below.

```
Phylogenetic analysis using the Neighbor-Joining method.
Genetic distance: F84
Input order randomized: No.
Outgroup: Annelid
Distance matrix used:

7
Chelicer1
Chelicer2  0.2238
Branchipod 0.2967 0.3374
```

```
Myriapod   0.2821 0.3160 0.2838
Annelid    0.3244 0.4054 0.3505 0.2997
Hexapod    0.2923 0.3084 0.2850 0.2440 0.3081
Malacostra 0.3734 0.4136 0.3705 0.3999 0.3946 0.3584


Pair-wise s/v ratio:

7
Chelicer1
Chelicer2  1.1191
Branchipod 0.4774 0.4318
Myriapod   0.6546 0.5386 0.7115
Annelid    0.5154 0.5918 0.8221 0.8847
Hexapod    0.4307 0.3291 0.7877 0.6536 0.7809
Malacostra 0.5828 0.4642 0.5743 0.6374 0.8215 0.5787

(Malacostra:0.2222,(((Chelicer1:0.0907,Chelicer2:0.1331):0.055
1,Branchipod:0.15):0.0017,(Myriapod:0.1243,Hexapod:0.1197):0.0
098):0.0123,Annelid:0.1724);
```

You see that phylogenetic reconstruction is very easy by using DAMBE. Below I explain various options you can choose before clicking the **Done** button. To understand the concepts involved in tree construction is much more difficult than to click a few mouse buttons.

2.1.1.1 Select phylogenetic algorithm

You can choose either the neighbor-joining or the Fitch-Margoliash method. The former is much faster than the latter, but the latter has a well specified optimization criterion, i.e., the least sum of squared differences between the observed and estimated genetic distances. Besides, you can do global optimization and input user trees with the Fitch-Margoliash method, but not with the neighbor-joining method. User trees are essential for evaluating relative statistical support for alternative phylogenetic hypotheses.

The default is the neighbor-joining method. If you click the option button for the Fitch-Margoliash method, then additional checkboxes (**Global optimization** and **User tree**) will be displayed. These checkboxes will disappear if you click the option button for the neighbor-joining method.

2.1.1.2 Select a genetic distance

There are many genetic distances that you can choose for using with either one of the two algorithms (Table 1). The simpler distances such as P, PoissonP, and JC69 are implemented only for teaching purposes and are rarely used in practical research. Please refer to the chapter dealing with genetic distances for more information on these distances.

Table 1. Genetic distances implemented in DAMBE. Distances based on Li (1993) will appear in the dropdown menu only when nucleotide sequences are protein-coding and when you click **Phylogenetics|Distance method|Codon-based distances**. Not listed are a number of amino acid-based distances incorporating various combination of physico-chemical properties.

Menu Item	Description
P	Proportion difference
PoissonP	P with Poisson correction for multiple hits
JC69	Jukes and Cantor's (1969) one-parameter distance
K2	Kimura's (1980) two-parameter distance
TN84	Tamura and Nei's (1984) distance
F84	Distance based on a substitution model in DNAML since 1984
TN93	Tamura and Nei's (1993) distance
Lake94	Lake's (1994) paralinear distance
Li93Ks	Synonymous substitutions (Li 1993)
Li93Ka	Nonsynonymous substitutions (Li 1993)
Li93Weighted	Weighted Ks and Ka

2.1.1.3 Choose an outgroup

All the OTUs in the sequence file are listed in the **Choose outgroup** dropdown menu. You can select any one of them by clicking the dropdown arrow. Someone has asked the possibility of forcing more than one outgroup species. I do not think this a good idea. If you are sure that species 1 and 2 should be outgroups coming out from the deepest node, it is a good idea to use either species 1 or species 2 as an outgroup and see if your sequence data can correctly put the other species close to the root. This can serve as a rough method of quality control. If species 1 and 2 end up in different lineages in your phylogenetic tree, and if you are absolutely sure that they should both come out from the deepest node, then your sequences must be poor. In other words, your sequences has failed to recover the true tree.

2.1.1.4 Choose other options

If you click the **Resampling statistics** checkbox, a dialog box will prompt you for how many data sets you wish to sample from the original data set. The default is 100. Two option button will also be displayed for you to choose either **Bootstrap** or **Jackknife** resampling method. If you have made relevant selections, clicking the **Done** button will produce a tree with bootstrap or jackknife values that we often encounter in literature. These resampling techniques are used both for the purpose of data validation and statistical tests. More details on bootstrap and jackknife are presented in later chapters on testing alternative phylogenetic hypotheses.

Click the **Randomize Input Order** checkbox will do the phylogenetic analysis repeatedly with randomized input order of OTUs.

Click the **Global optimization** check box, which appears only when the Fitch-Margoliash method is selected, will do the following. After the last species is added to the tree, each possible clade is removed and re-added to the tree. This improves the result, since every species is reconsidered.

If you click the **User tree** checkbox, the three checkboxes mentioned above will disappear, and two new option buttons, **Trees from a file** and **All possible trees**, appear. Note that the number of possible rooted and unrooted trees, designated by N_R and N_U, respectively, increases rapidly with increasing number of species according the following equations:

$$N_U = \frac{(2n-5)!}{2^{n-3}(n-3)!} \tag{19.3}$$

$$N_R = \frac{(2n-3)!}{2^{n-2}(n-2)!} \tag{19.4}$$

There are 2,027,025 rooted trees for only 9 species, so you should exert caution when clicking the **All possible trees** option button. If you are specialized in some particular groups of organisms, you generally know it is unnecessary to consider all possible topologies. For example, if you have a DNA sequence from orthologous genes from two identical twins and two other unrelated individuals, it is unnecessary to consider the possibility of the two identical twins each clustering with one of the two unrelated individuals. It is generally true that phylogenetic controversies can be formulated into just a few alternative phylogenetic hypotheses. These hypotheses can be written down as user trees in PHYLIP format, which can be read into DAMBE for evaluation of relative statistical support. The user trees should be all **unrooted**. DAMBE will check the first tree. If it is rooted, you will get a warning. If it is unrooted, DAMBE will assume that all the rest of trees will be the same, i.e., unrooted. Do not check the **User tree** checkbox for the time being because the use of user trees will be dealt with fully in later chapters on testing alternative phylogenetic hypotheses.

2.1.2 DAMBE and distance methods based on allele frequencies

A file containing allele frequency data is typically in the following PHYLIP format:

```
5   4
```

```
2 2 3 2
European   0.7285 0.6386 0.0205 0.8055 0.5043
African    0.9675 0.9511 0.0600 0.7582 0.6207
Chinese    0.7986 0.7782 0.0726 0.7482 0.7334
American   0.8603 0.7924 0.0000 0.8086 0.8636
Australian 0.9000 0.9837 0.0396 0.9097 0.2976
```

The two numbers on the first line indicates, respectively, the number of OTUs and the number of loci. The second line shows the number of alleles per locus, i.e., the first, second and forth locus each have 2 alleles, and the third locus has 3 alleles. The next five data lines show the allele frequencies of the five loci, typically with one allele omitted for each locus. For example, one allele for first locus has a frequency of 0.7285, and the frequency of the other allele is simply 1 – 0.7285 and is omitted. Similarly, the first two alleles of the third locus have frequencies 0.0205 and 0.8055, respectively, and the frequency of the third allele is simply one minus these two frequencies and is omitted.

To use allele frequency data for phylogenetic reconstruction using DAMBE, click **File|Open**. A standard File/Open dialog appears. In the **Files of type** dropdown menu, click the last item **Allele frequency files (*.FRE)**. All files with the extension .FRE will be displayed. Double-click a file will read the data into DAMBE. Or you can click the file once to highlight it and then click the **Open** button.

A dialog box is displayed (fig. 5). The two option buttons at the top is for you to specify your file format, i.e., whether the input data file contains all gene frequencies or with one allele omitted at each locus. The three different genetic distances based on allele frequency data have been presented in the chapter dealing with genetic distances, which should be consulted if you are to use allele frequency data for phylogenetic reconstruction. All other options are the same as in the previous section, and the user tree should be **unrooted**. DAMBE will check the first tree. If it is rooted, you will be issued a warning. The **User tree** checkbox appears only when the **Fitch-Margoliash** method is chosen. Do not check the **User tree** checkbox for the time being because the use of user trees will be dealt with fully in the next chapter on testing alternative phylogenetic hypotheses. Click the **Go!** button and a phylogenetic tree will be generated.

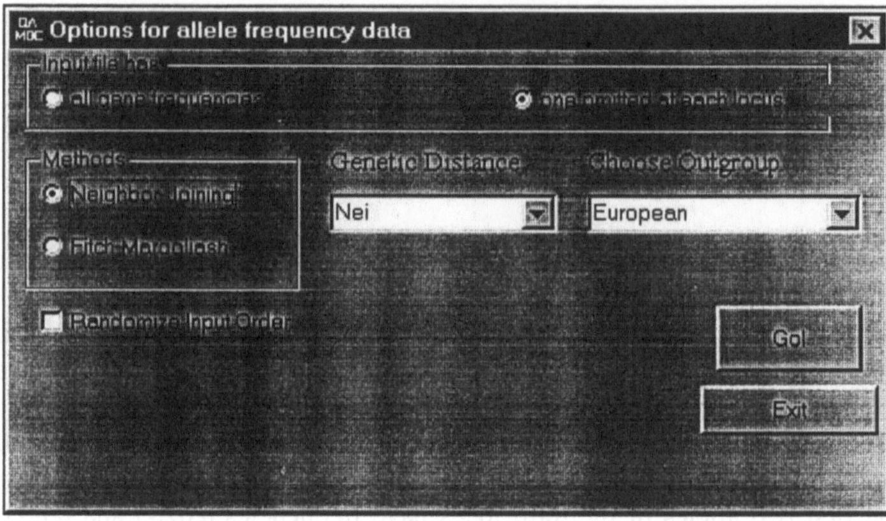

Figure 5. Dialog box for phylogenetic analysis using allele frequency data. More checkboxes and option buttons will appear in response to your mouse clicks.

2.2 Maximum parsimony method

The maximum parsimony method can be viewed as an approximation to the maximum likelihood method (Baldi and Brunak 1998, p. 226). It implicitly assumes that substitutions are rare (which implies that sequences need to be very long to have sufficient number of informative sites), that the substitutions are uniform over sites, and that the substitution rate is constant over time in different lineages. There are a number of cases in which the maximum parsimony approach would be positively misleading (Felsenstein 1978; Nei 1991; Takezaki and Nei 1994). Because of these assumptions, the maximum parsimony approach is not good for reconstructing ancient phylogenies.

The parsimony algorithm implemented in DAMBE is the same as the DNAPARS program in PHYLIP. In fact I copied almost all codes in DNAPARS into my **mktree.dll** program with little modification. For this reason, the reader is referred to the documentation for DNAPARS in PHYLIP for further information. This section provides a brief introduction on how to do a maximum parsimony reconstruction by using DAMBE.

Start DAMBE and open a file with nucleotide sequences, e.g., the **invert.fas** file that comes with DAMBE. Click **Phylogenetics|DNAMP** and a dialog box (fig. 6) appears for you to select options available for doing a maximum parsimony analysis.

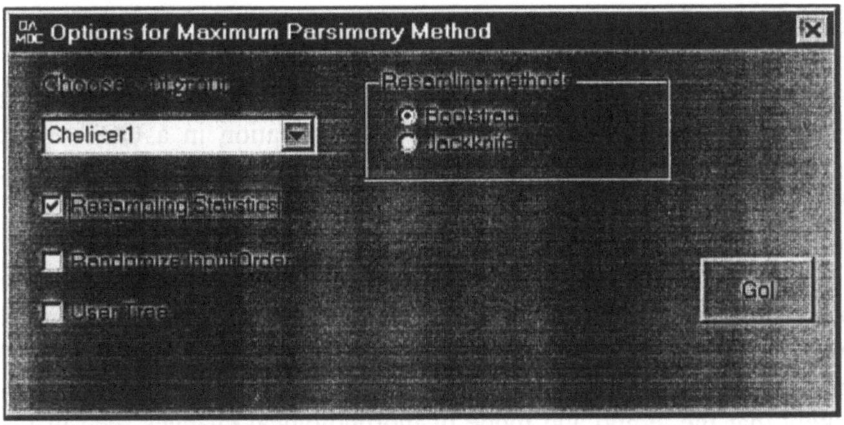

Figure 6. Dialog box for phylogenetic analysis using maximum parsimony.

All the options are the same as those shown in fig. 3, except that the user tree should now be **rooted**, in contrast to the unrooted user tree used for the Fitch-Margoliash method. DAMBE will check the first tree in the tree file. If it is unrooted, you will be issued a warning. Click the **Go!** button will generate a tree, with no branch lengths. The tree display window will display the tree with equal branch lengths, just to remind you that there is no branch length estimate.

Just as in the Fitch-Margoliash method, the **User Tree** checkbox can be used for evaluating relative statistical support for alternative topologies. This will be dealt with in later chapters.

2.3 Maximum likelihood method

2.3.1 Preamble

The substitution models assumed by various implementations of the maximum likelihood method in phylogenetics share the following assumptions: (1) the substitutions occur independently in different lineages; (2) substitutions occur independently among sites; (3) the process of substitution is described by a time-homogeneous Markov process, and (4) the process of substitution is stationary. It is important to realize at the very beginning that the validity of maximum likelihood reconstruction depends much on how close its assumed substitution model approximates reality.

Except for the first assumption, assumptions 2, 3, and 4 are often false. The second assumption, that substitutions occur independently among sites, is obviously not true. Take two codons, GAT and GGT, for example. Both codons end with a T. Whether a T→A substitution would occur depends much on whether the second position is an A or a G. The T→A substitution is rare when the second codon position is A because a T→A mutation in the GAT codon is nonsynonymous, but relatively frequent when the second codon position is G because such a T→A mutation in a GGT codon is synonymous. This violates the assumption that nucleotide substitutions occur independently among sites. One can easily come up with many other examples in which the assumption does not hold.

The third assumption, that the process of substitution is described by a time-homogeneous Markov process, is also unlikely to be true. The recent proposal of punctuated equilibrium (Eldredge 1989) based on fossil records suggests that evolution occurs very sporadically, rather than continuously. It is likely that the tempo and mode of morphological changes seen in fossils do not have much relevance to molecular evolution. However, a compilation of molecular evidence appears to suggest sporadic evolution at the molecular level as well (Gillespie 1991).

The fourth assumption, that the process of substitution is stationary, is also problematic. Suppose we wish to reconstruct a tree from a group of orthologous sequences from both invertebrate and vertebrate species. There is little DNA methylation in invertebrate genomes, but heavy DNA methylation in some vertebrate genomes. DNA methylation greatly enhanced the C→T transition (and consequently the G→A transition on the opposite strand. The net result is a much elevated transition/transversion bias in the lineages with DNA methylation, as well as an increased nucleotide frequencies for nucleotides A and T. This substitution process is clearly not stationary. If your sequences differ much in nucleotide frequencies, then you should use a distance method with the paralinear (Lake 1994) or LogDet (Lockhart et al. 1994) distance which is based on a substitution model that does not assume stationarity.

2.3.2 Phylogenetic reconstruction with DAMBE

The maximum likelihood algorithm implemented in DAMBE is the same as the BASEML program in the PAML package (Yang 1997b). I copied almost all the code in the BASEML program into my **mktree3.dll** file with little modification. Consequently, the output is almost exactly the same as the output from the BASEML program in PAML. The BASEML program has rather limited tree-searching algorithms, but features a rich implementation of substitution models. It is best not at finding the best tree, but at testing phylogenetic hypotheses.

Start DAMBE and open a file containing aligned nucleotide sequences. Click **Phylogenetics|Maximum Likelihood|Nucleotide Sequences**. A dialog box appears (fig. 7). In the **Run Mode** dropdown menu, you can choose either **Search Best Tree**, or **Semi-automatic** (for resolving a partially resolved tree) or **User tree**. Do not check the **User tree** checkbox for the time being because the use of user trees will be dealt with fully in the later chapter on reconstruction of ancestral sequences and on testing alternative phylogenetic hypotheses.

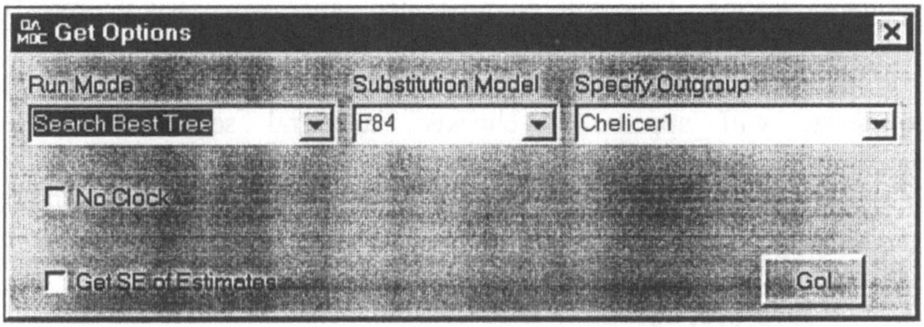

Figure 7. Dialog box for phylogenetic analysis using the maximum likelihood method.

There are three nucleotide-based substitution models available: the F84 model in the DNAML program in the PHYLIP package since 1984 (Felsenstein 1993), the HKY85 model (Hasegawa et al. 1985), and the TN93 model (Tamura and Nei 1993). Simpler models, such as JC69 and K2 models, are not included because they are known to be unrealistic.

It cannot be emphasized enough that the validity of phylogenetic analyses using the maximum likelihood method depends on how close the underlying substitution model can approximate reality. A substitution model typically has two categories of parameters for describing a substitution pattern (Kumar et al. 1993; Li 1997; Yang 1997b), i.e., the rate ratio parameters (often symbolized by α, β, δ, γ, etc.) and the frequency parameters, i.e., π_A, π_C, π_G, and π_T. The JC69 and K80 models assume equal frequency parameters. The number of rate ratio parameters for JC69, K80, F84, HKY85 and TN93 models are 0, 1, 1, 1, and 2, respectively.

The F84 and HKY85 models are the simplest models that include both frequency parameters as well as a rate ratio parameter to accommodate the s/v bias. It has been found that transitions are also heterogeneous, with A↔G transitions having a different rate from T↔C transitions (Tamura and Nei 1993), and the TN93 model is proposed to accommodate this rate heterogeneity among transitions.

The **Clock** checkbox, when checked, will impose a molecular clock. A phylogenetic tree with a molecular clock can be used to date speciation events, but you should remember to test the molecular clock hypothesis first. Details of testing the molecular clock hypothesis is presented in later chapters. When the run mode is **User tree**, then the **Clock** checkbox will disappear. Whether a clock is assumed depends on whether the user trees are rooted or unrooted, with a rooted tree imposing a molecular clock. Do not mix rooted and unrooted trees in the same tree file.

If the **Get SE of Estimates** is checked when the run mode is set to **User tree**, then the standard error (SE) of estimated parameters will be calculated as the square roots of the large sample variances. These SE's can be used for hypothesis testing. This will be illustrated latter when we interpret the output.

If the run mode is **User tree**, another checkbox, **Get ancestral sequences**, will appear. If checked, ancestral sequences will be reconstructed, and stored in a file together with original sequences. Such a file will have a **.RST** file extension by default to avoid confusion with other types of sequence files.

2.3.3 Interpret output

The output below is produced by choosing run mode **Search for best tree**, and the **F84** model, and have the **Clock** checkbox unchecked. I used only five sequences in the file to save space. The output is slightly modified so that it will fit the page size. The output is of two parts, one being a phylogenetic tree shown is DAMBE's tree display window, and the other being text output shown DAMBE's text display window. Only the latter part is duplicated below:

```
Frequencies..
                          T      C      A      G
Chelicerl           0.2340 0.2461 0.2664 0.2535
Branchipod          0.2729 0.2350 0.2627 0.2294
Annelid             0.2368 0.2516 0.2525 0.2590
Hexapod             0.2683 0.2183 0.2599 0.2535
Malacostra          0.2683 0.2192 0.3071 0.2054

Average             0.2561 0.2340 0.2698 0.2401
```

Check the nucleotide frequencies to make sure that they are similar, i.e., conform to the assumption of stationarity. The average values are often used as empirical estimates of frequency parameters π_T, π_C, π_A, and π_G.

```
Distances:  F84(kappa)  (alpha set at 0.00)
This matrix is not used in later m.l. analysis.

Chelicerl
Branchipod    0.2917( 0.4804)
```

```
Annelid        0.3183( 0.4728)  0.3440( 0.7767)
Hexapod        0.2886( 0.4086)  0.2799( 0.8004)  ... ...
Malacostra     0.3712( 0.5587)  0.3669( 0.5734)  ... ...
```

"alpha" above refers to the α parameter in a gamma distribution. Its value, 0.00, means that rate heterogeneity over sites is ignored. This results in underestimates of genetic distances, but avoids inflation of bias.

```
stage 1:    10 trees
star tree: (12345) lnL:  -4739.762988
(Chelicer1:0.16290, Branchipod:0.16483, Annelid:0.20313,
Hexapod:0.15752, Malacostra:0.26318);
```

Tree reconstruction begins with a star tree. We pick up all possible pairs from the five OTUs, group them together, and see which resulting topology has the largest likelihood. There are $C_5^2 = 10$ combinations.

```
S=1:   1/  10  T=   1 ((12)345)
lnL(ntime:  6  np:  7):  -4729.624841
 6..7      7..1      7..2      6..3      6..4      6..5 ← Branch
0.03243  0.14945  0.15121  0.19369  0.14682  0.25826  0.62650

S=1:   2/  10  T=   2 ((13)245)
lnL(ntime:  6  np:  7):  -4729.268886
 6..7      7..1      7..3      6..2      6..4      6..5
0.03490  0.14369  0.18607  0.15661  0.14817  0.26033  0.60620

...... (omitted)

S=1:   8/  10  T=   8 (12(34)5)
lnL(ntime:  6  np:  7):  -4714.936358 ← Largest among the 10.
 6..1      6..2      6..7      7..3      7..4      6..5
0.15286  0.15326  0.05191  0.17524  0.12767  0.25146  0.60328

S=1:   9/  10  T=   9 (12(35)4)
lnL(ntime:  6  np:  7):  -4717.042787
 6..1      6..2      6..7      7..3      7..5      6..4
0.16001  0.15751  0.05608  0.16142  0.22544  0.14926  0.61949

S=1:  10/  10  T=  10 (123(45))
lnL(ntime:  6  np:  7):  -4724.046877
 6..1      6..2      6..3      6..7      7..4      7..5
0.15657  0.15967  0.19824  0.04201  0.12621  0.23675  0.60958
```

The eighth topology has the largest likelihood and is chosen as the best partially resolved tree for the next stage. This algorithm is called star decomposition. The numbers under the branches are branch lengths, except for the last value which is the estimated κ which measures transition/transversion bias. Note that κ (=0.60958) under the F84 model is NOT the conventional ratio of α/β, which can be obtained by using the following equation:

$$\frac{\alpha}{\beta} = 1 + \frac{\left(\frac{\pi_T \pi_C}{\pi_T + \pi_C} + \frac{\pi_A \pi_G}{\pi_A + \pi_G}\right)\kappa_{F84}}{\pi_T \pi_C + \pi_A \pi_G} = \kappa_{HKY85} \tag{19.5}$$

where π values are frequency parameters. You may use the four nucleotide frequencies averaged over the five sequences as estimates of π values and obtain a α/β ratio of about 2.

As the equation has shown, the κ value under the HKY85 model is the conventional ratio of α/β, which makes it easy to test whether there is transition bias. If you set run mode to **User tree** and check the **Get SE** checkbox, then the standard error of κ will be calculated for you to compute the z-score:

$$z = \frac{\kappa - 1}{SE_\kappa} \tag{19.6}$$

If z is > 1.96, then there is significant transition bias at the 0.05 level. Under the TN93 model, there will be two κ values, κ_1 for T↔C transitions and κ_2 for A↔G transitions. If you wish to test the null hypothesis of $\kappa_1 = \kappa_2$, the z score is computed as

$$z = \frac{\kappa_1 - \kappa_2}{\sqrt{SE^2_{\kappa_1} + SE^2_{\kappa_2}}} \tag{19.7}$$

which, if larger than 1.96, can be declared as significant at the 0.05 level. Note that the denominator, the summation of the two variances, implies that there is no covariance between κ_1 and κ_2. A better test is the likelihood ratio test, carried out as follows. You use the same tree to obtain the log-likelihood for the two models, e.g., HKY85 and TN93, $\ln L_{HKY85}$ and $\ln L_{TN93}$. We then calculate $X^2 = 2*(\ln L_{TN93} - \ln L_{HKY85})$, which can be examined against chi-square distribution with one degree of freedom (because Tn93 model has one more parameter than the HKY85 model). If X^2 is larger than 3.84, then we can declare that the two κ values are significantly different at 0.05 level.

```
stage 2:       3 trees
star tree: (12(34)5) lnL:   -4714.936358
(Chelicer1:0.15286, Branchipod:0.15326, (Annelid:0.17524,
Hexapod:0.12767):0.05191, Malacostra:0.25146);

S=2:    1/    3   T=   11 ((12)(34)5)
lnL(ntime:   7  np:   8):   -4712.263615
 6..8     8..1     8..2     6..7     7..3     7..4     6..5
0.0157  0.1480  0.1486  0.0452  0.1755  0.1271  0.2439  0.6102
```

```
S=2:   2/   3  T=  12 ((1(34))25)
lnL(ntime:  7  np:  8):  -4706.425786
 6..8     8..1     8..7     7..3     7..4     6..2     6..5
0.0309  0.1444  0.0448  0.1756  0.1269  0.1345  0.2364  0.6041

S=2:   3/   3  T=  13 ((15)2(34))
lnL(ntime:  7  np:  8):  -4707.432003
 6..8     8..1     8..5     6..2     6..7     7..3     7..4
0.0289 0.1361  0.2386  0.1447  0.0457  0.17877  0.1244  0.6214

best tree: ((1(34))25)    lnL:  -4706.425786
((Chelicer1:0.14443, (Annelid:0.17558,
Hexapod:0.12688):0.04480):0.03091, Branchipod:0.13449,
Malacostra:0.23643);
```

The resulting tree is displayed in DAMBE's tree display window.

2.4 Reconstructing Ancestral Sequences

There are two situations where reconstructed ancestral sequences may be useful. First, when one wishes to estimate parameters (e.g., rate ratio parameters, the shape parameter in a gamma distribution, transition/transversion ratio, etc.) in various substitution patterns, one typically needs substitution data obtained from pair-wise comparisons between neighboring nodes along a phylogenetic tree. This procedure requires ancestral states of internal nodes. Second, ancestral sequences are needed when one wants to infer which nucleotide or amino acid sites have experienced which kind of substitution during which evolutionary period.

DAMBE uses codes in the BASEML program (Yang 1997b; Yang et al. 1995) for reconstructing ancestral sequences. You need a tree topology in PHYLIP format as well as an aligned sequence file, e.g., the **invert.fas** file that comes with DAMBE, for reconstructing ancestral sequences. If you do not have a file with a tree, then just use one of the phylogenetic methods we have just covered to obtain a tree and save it in a file. For ease of presentation, let us assume that you have already obtained such a tree and saved it in a file named **invert.dnd**.

Start DAMBE and open a sequence file, e.g., the **invert.fas** file that comes with DAMBE. Once the sequences are displayed in DAMBE, click **Phylogenetics|Maximum Likelihood|Nucleotide sequences**. A dialog box (fig. 7) appears for you to specify options. Under **Run Mode**, choose **User tree**. A standard **File/Open** dialog box will appear for you to specify which file contains the user tree. Click the file, e.g., **invert.dnd**, that contains the tree for the seven invertebrate species and then click the **Open** button. Check the **Get ancestral sequences** checkbox and click the **Go!** button. After waiting for one minute or two, you will be prompted to enter a file name to save the reconstructed sequences. The file will then be saved in text format, and the ancestral sequences will be added to the existing sequences for you to apply further pair-wise comparisons between neighboring nodes along the tree. For example, if you click **Seq. Analysis|Nucleotide difference|Detailed output**, only sequence pairs that are neighbors along the phylogenetic tree will be shown in the ensuing dialog box. If you do not have ancestral sequences, then all possible sequence pairs will be shown in the dialog box.

After reconstructing the ancestral sequences, DAMBE will disable the **Phylogenetics** menu. There are two reasons for this. First, it makes little sense to perform phylogenetic reconstruction on reconstructed ancestral sequences. Second, some reconstructed ancestral sequences will be the same as the existing sequences, but some phylogenetic functions in DAMBE assume that all sequences are different from each other.

Now that we have learned how to reconstruct phylogenetic trees by using the distance, maximum parsimony and maximum likelihood methods, it is time for your read a few comparative studies that assess the performance of different phylogenetic methods by simulation (Felsenstein 1988b; Kuhner and Felsenstein 1994; Nei 1991). Results from these simulation studies are not always consistent. Below I will just highlight five thorny problems that a practising phylogeneticist should pay attention to. The first two concern the quality of the sequence data, and the last three concerns phylogenetic algorithms.

The first problem that is often neglected concerns sequence alignment, which is the beginning of all comparative sequence analyses. If we cannot obtain reliable sequence alignment, then all subsequent analyses will be nonsense. In particular, alignment of rRNA and tRNA sequences is almost always a nightmare, especially when the sequences have diverged over a long time. For this reason, it is often necessary to incorporate the secondary structure of the RNA molecules in the multiple sequence alignment (Dixon and Hillis 1993; Gutell et al. 1990; James et al. 1989; Pace et al. 1989; Srikantha et al. 1994).

The second problem is substitution saturation. Substitution saturation decreases phylogenetic information contained in the sequences, and has plagued the phylogenetic analysis involving deep branches, such as major arthropod groups. When sequences have reached full substitution saturation, then the genetic distances among the sequences will depend entirely on nucleotide frequencies among sequences, and nucleotide frequencies are generally not phylogenetic indicators.

There are currently two main approaches for finding whether molecular sequences contain phylogenetic information. The first approach involves the randomization test or permutation test (Archie 1989; Faith 1991), the second employs the standard g_1 statistic for measuring the skewness of branch lengths of alternative trees (Hillis and Huelsenbeck 1992). Both approaches suffer from the problem that, as long as we have two closely related species, the tests will lead us to conclude the presence of significant phylogenetic information in the data set even if all the other sequences have experienced full substitution saturation. Besides, these methods do not measure substitution saturation directly, i.e., they only assess one of the possible consequences of substitution saturation. DAMBE provides an alternative measure of sequence saturation based on entropy, implemented under the **Seq. Analysis|Measure substitution saturation**.

The third problem arises from the violation of the molecular clock assumption. Some lineages occasionally would evolve much faster than their sister lineages, and create what is known as the long-branch attraction problem. This is nicely illustrated by a simple data set in Swofford et al. (1996). The maximum likelihood method has been shown to be very robust against nonconstant substitution rates among different lineages.

The four problem is the rate heterogeneity over sites, which defeats virtually all methods currently available for phylogenetic reconstruction (Kuhner and Felsenstein 1994). Although the use of a gamma distribution to accommodate the rate heterogeneity has been suggested to improve phylogenetic estimation in some particular situations, results from simulation studies are not encouraging (Yang 1997a).

The fifth problem occurs when the stationarity assumption of the substitution process is violated. I have already mentioned the problem and the solution in the section on maximum likelihood and will not repeat here.

3. EXERCISE

Open the **ape5.fas** file that comes with DAMBE. This file contains the cytochrome oxidase subunit I (COI) sequences from five primate species. Establish a phylogenetic tree using distance, maximum parsimony and maximum likelihood methods. For distance methods (neighbor-joining and Fitch-Margoliash), use different genetic distances and decide which genetic distance is more appropriate. Based on your exploration, discuss the advantage and disadvantage of different phylogenetic methods. Save a rooted tree to a file named **ape5r.dnd** and an unrooted tree to a file named **ape5ur.dnd**.

You must have read news stories about the mitochondrial Eve. The dating of when the mitochondrial Eve lived assumes a valid molecular clock, i.e., the mitochondrial genes have evolved at similar rates in different human lineage. The two tree files you have just saved will be used in later chapters for testing whether the molecular clock represented by the COI gene is constant or not in different primate lineages.

Chapter 20

Testing the Molecular Clock Hypothesis

In previous chapters we have learned the concept of the molecular clock and the fundamentals of phylogenetic reconstruction. We used the molecular clock without checking its validity, and we reconstructed phylogenetic trees without evaluating their statistical support relative to alternative trees. This chapter introduces a method for testing the validity of the molecular clock hypothesis. Evaluating relative statistical support for alternative phylogenetic hypotheses is treated in a later chapter.

Evolutionary biologists often use a molecular clock to date speciation events, or infer when and where the common ancestor of a population has lived. For example, a few years ago, it has been claimed that the mitochondrial Eve lived somewhere in Africa about 200,000 years ago (Cann et al. 1987). This is based on the assumption that mitochondrial clock is constant in different human lineages. Is this a valid assumption? In the previous chapter on molecular phylogenetics, we have already mentioned several factors causing the molecular clock to tick at different rates in different lineages, i.e., the generation time effect (Ellsworth et al. 1993; Gu and Li 1992; Seino et al. 1992), the germ line effect (Miyata et al. 1987) and the effect of genetic context (see the chapter on molecular phylogenetics for more details). Will the molecular clock still works in spite of all these distorting factors?

The molecular clock hypothesis is typically formulated as the null hypothesis of equal evolution rates along different lineages, which can be tested by using the likelihood ratio test implemented in DAMBE. A likelihood ratio test is a significance test. We will first refresh your memory of basic concepts of statistical significance tests by reviewing the simple t-test. The t-test is then cast in the framework of a likelihood ratio test. The only purpose of presenting this simple t-test is just to let you see the similarity between a simple t-test and a more involved likelihood ratio test so that you will be more courageous in future statistical endeavours. The last part of the chapter deals with how to use DAMBE to carry out a likelihood ratio test of the molecular clock hypothesis.

1. THE T-TEST AND ITS APPLICATIONS

Suppose we have a normally distributed variable x with sample values x_1, x_2, ... x_n, and we are interested in whether the mean value of x is significantly different from x_0. This is a typical situation for a t-test, with the null hypothesis being $\bar{x} = x_0$. The test is done as follows:

$$\bar{x} = \frac{\sum_{i=1}^{n} x_i}{n}$$

$$s^2 = \frac{\sum_{i=1}^{n} (x_i - \bar{x})^2}{n-1}$$

$$s_{\bar{x}} = \sqrt{\frac{s^2}{n}} \tag{20.1}$$

$$t = \frac{\bar{x} - x_0}{s_{\bar{x}}}$$

We see that if $\bar{x} = x_0$, then t = 0. If $\bar{x}$ is very different from x_0, then the absolute value of t will be large and the likelihood of the null hypothesis being true becomes smaller. Thus, we can view the absolute value of the t statistic as a measure of the difference between the observed value and the expected value under the null hypothesis. A large t means a great deviation of the observation from the expectation. If the absolute value of t is larger than a predetermined value, then we reject the null hypothesis.

The t statistic has a probability distribution called t distribution, which is similar to the normal distribution except that it is flatter and with longer tails. If the sample size is infinitely large, then the t distribution converges to the normal distribution. The t-test is invented for small samples from which the estimated mean and variance have a small chance of being identical to the population mean and variance. It is necessary to know the distribution of the test statistic in order to carry out a valid statistical significance test. If we do not know the t distribution, then the t statistic will be no more than an index of differences between the observation and the expectation.

You might have learned the story about how Charles Darwin was frustrated about his inability to reach a statistical conclusion from his breeding data (Peters 1987, pp. 110-111). Darwin wanted to know the effect of outcrossing on the effect of plant size. So he chose 15 plant species for an experiment. For each species, he forced some individuals to inbreed and some to outbreed. He measured the growth characters for the inbred individuals and the outbred ones. So he obtained 15 pairs of data over 11 years. When he presented his hard-earned data to his cousin, Francis Galton, he was disappointed to be told that much more data were needed for him "to be in a position to deduce fair results." Galton was not able to deduce fair results from Darwin's data because the t distribution is unknown at that time.

In summary, to make a significance test, we need to have two things. First, we need a test statistic, such as t, that measures the difference between the observed value and the expected value. Second, we need to know the probability distribution of the test statistic, such as the t distribution, from which we can draw probability statement. Below we look at what test statistic is used in a likelihood ratio test and what probability distribution is used to draw probability statement.

2. THE LIKELIHOOD RATIO TEST

Like the t-test, the likelihood ratio test is also a significance test. It tests whether a simpler model fits the observed data as well as a more general model, where the simpler model is a special case of the more general model. For example, the simple model could be $y = x$, and the general model could be $y = x^b$. You see that the simple model is a special case of the general model when $b = 1$. Obviously, the general model will fit the observed data at least as good as the simple model.

In the example above involving the t-test, the simple model sets $\bar{x} = x_0$, which is the null hypothesis. The alternative hypothesis is $\bar{x} \neq x_0$. The general model is the union of the null hypothesis and the alternative hypothesis, i.e., $\bar{x} = x'$, where x' can take any real value including x_0. In a likelihood ratio test, the set of values a parameter (e.g., $\bar{x}$) can take given the null hypothesis is often designated by Ω_0. In our example of t-test, Ω_0 contains a single value of x_0. The set of values a parameter can take given the alternative hypothesis is designated by Ω_a, which is any value other than x_0 in our example of t-test. The general model has $\Omega = \Omega_0 \cup \Omega_a$.

In the t-test, we calculate the t statistic. In a likelihood ratio test, we calculate a λ statistic which is defined as

$$\lambda = \frac{L(\Omega_0)}{L(\Omega)} \tag{20.2}$$

where $L(\Omega_0)$ and $L(\Omega)$ denote the likelihood functions for the simple model and the general model, respectively. Given the set of sample values x_1, x_2, ..., x_n for the normally distributed random variable x, we can write the likelihood function as

$$L = \frac{e^{\frac{-(x_1-\mu)^2}{2\sigma^2}}}{\sigma\sqrt{2\pi}} \cdot \frac{e^{\frac{-(x_2-\mu)^2}{2\sigma^2}}}{\sigma\sqrt{2\pi}} \cdot \ldots \cdot \frac{e^{\frac{-(x_n-\mu)^2}{2\sigma^2}}}{\sigma\sqrt{2\pi}}$$

$$= \frac{e^{\frac{-\sum_{i=1}^{n}(x_n-\mu)^2}{2\sigma^2}}}{\sigma^n(2\pi)^{n/2}} \tag{20.3}$$

$L(\Omega_0)$ is obtained by replacing μ by x_0, and σ^2 by s_0^2, which is calculated the same way as s^2 in equation except that $\bar{x}$ is replaced by x_0. Similarly, $L(\Omega)$ is obtained by replacing μ by $\bar{x}$, and σ^2 by s^2. With some algebraic manipulation, it can be shown that

$$\lambda = \frac{L(\Omega_0)}{L(\Omega)} = \left(\frac{s^2}{s_0^2}\right)^{n/2} = \left(\frac{n-1}{t^2+n-1}\right)^{n/2} \tag{20.4}$$

where t is the t statistic in equation (20.1). Recall that we have viewed the absolute value of the t statistic as a measure of the difference between the observed value and the expected value (expected under the null hypothesis), with $t = 0$ implying a perfect match between the observed and the expected value. The statistic λ is the opposite of t, and reaches the maximum value of 1 when $t = 0$. A value of λ close to zero means that the likelihood of the null hypothesis being true is very small.

In a t-test, the difference between $\bar{x}$ and x_0 is declared as significant when the absolute t value is larger than a predetermined value such as $t_{\alpha,df}$. Similarly, the λ value in a likelihood ratio test is declared as significant if it is smaller than a predetermined value. For the normal distribution and large n, $-2\ln(\lambda)$ has approximately a χ^2 distribution with degrees of freedom equal to N - N_0, where N and N_0 are, respectively, the number of parameters to be estimated in the general and the specific model. In the t-test example, the general model estimates $\bar{x}$, whereas the specific hypothesis assumes $\bar{x} = x_0$. So N - N_0 = 1.

Equation (20.4) shows that the λ statistic and the t statistic have one to one correspondence, i.e., the likelihood ratio test employing the λ statistic can be translated into a t-test using the t statistic. Alternatively, a t-test can be transformed into a likelihood ratio test. In short, the likelihood ratio test and the t-test presented above are identical tests.

Just like the t-test, the corresponding likelihood ratio test is a parametric test and has assumptions concerning the distribution of the x variable. We can do a t-test without checking its assumptions, such as normality and, in the two-sample t-test, homoscedasticity (equal variance), but the result of the significance test may be misleading. Similarly, we can do a likelihood ratio test without checking its assumptions, producing potentially misleading results.

One illustrative example may help you appreciate this point. Suppose you have a variable x with four sampling values, 1, 2, 2, and 3, and you are interested in if the mean of x is significantly different from zero. You can carry out a t-test or a likelihood ratio test by assuming a normal distribution, and the significance test will show that $\bar{x}$ (= 2) is significantly different from zero, with p = 0.0163. Suppose we take one more observation and obtain a value of 10. Now $\bar{x}$ becomes 3.6, much larger than the previous mean of 2. Intuitively, we would expect the new $\bar{x}$ to be even more significantly different from zero. However, if you do a t-test or a likelihood ratio test, you will find that the new $\bar{x}$ is no longer significantly different from zero, with p = 0.1144.

The counter-intuitive result arose because the variable x may not be normally distributed. The five values, including the new value of 10, might well be from a right-skewed distribution, i.e., the normality assumption for the t-test or the equivalent likelihood ratio test is violated. It is often said that a likelihood-based method is robust, but robust to which extent? One can easily verify that a nonparametric test employing ranked data would be much better than the parametric tests given the five fictitious data points.

Likelihood-based methods are generally robust when the sample size is very large. In our t-test example, even if the underlying distribution is not normal, the distribution of the sample means will be approximately normal if the sample size is large. Given the increasingly longer sequences in modern comparative sequence analysis, one might argue that sample sizes are large in phylogenetic analysis. This is again a misconception. The sequence data are typically heterogeneous among sites, and this heterogeneity increases with sequence length. For example, when you have a set of short orthologous intron sequences, there might be little heterogeneity among sites. However, if you include the flanking exons to make sequences longer, the heterogeneity greatly increases. Pooling heterogeneous data together almost always leads to violation of assumptions underlying parametric methods.

Let me reiterate the point that statistics is best taught and written about by professional statisticians who can always explain statistical matters in a more terse and lucid manner than a biologist like me. For this reason you are strongly encouraged to learn from authoritative statistical books. I personally feel guilty for "smuggling" some not-so-good statistics to your hands.

3. TEST THE MOLECULAR CLOCK HYPOTHESIS

In testing a molecular clock, the simple model is one with constant evolutionary rate along different lineages, so that all extant species have the same genetic distance back to the common ancestor. The general model does not assume a constant rate and allows different lineages to have different branch lengths. For a tree with n OTUs, there are 2n - 3 time parameters to estimate without assuming the clock, but only n - 2 time parameters assuming the clock. Thus, in a likelihood ratio test of the molecular clock hypothesis, the degree of freedom is (2n - 3) - (n-2) = n - 1.

At the end of the chapter on molecular phylogenetics, you are asked to do an exercise by applying phylogenetic analysis to a sequence data file, **Ape5.fas,** containing the mitochondrial COI gene from five primate species. We will use the same sequence file and the phylogenetic tree that you have derived from the sequences. A test of the molecular clock hypothesis should always be done with the same tree, otherwise it is meaningless. The following tree is from my own phylogenetic analysis of the sequences:

```
((((CHBCO1,HSMCO1),GGMCO1),ORACO1),HLMCO1);
```

You should have a file containing such a tree, with or without branch lengths, in plain text format. If your file contains multiple trees, then only the first tree will be used. It is always recommended to have a well established phylogenetic tree, preferably derived from alternative sources of data, for a serious test of the molecular clock hypothesis, because the test result is topology-dependent. For ease of presentation, let us assume that the name of your tree file is **Ape5.dnd**.

Click **Phylogenetics|Test Molecular Clock**, and you will be reminded of the requirement of a topology saved in a file. As you have already obtained such a file, just click the **Yes** button to proceed. A dialog box will then appear for you to choose which substitution model to use in the test. Choose the **F84** model (or any other model) and click the **Go!** button. A standard **File/Open** dialog box appears prompting you to choose your tree file. Click the **Ape5.dnd** file and then click the **Open** button. Wait for a few second for DAMBE to finish the test. You will again be prompted for a file name, this time a file for saving the test result. Enter whatever name you like, and the result will be saved as well as displayed in DAMBE's display window, reproduced below:

```
Testing the null hypothesis that different lineages
evolve at the same rate.

The topology used for the test:
((((CHBCO1,HSMCO1),GGMCO1),ORACO1),HLMCO1);

Relevant statistics:
=================================================
Substitution model:                         F84

Likelihood with no clock:            -4441.0482
Likelihood with  clock:              -4442.8850
Likelihood ratio chisquare:              3.6737
Degree of freedom:                       4
Prob.:                                   0.4520
=================================================
Note: Prob. above is the probability that you would be wrong
if you reject the null hypothesis that there is no difference
in evolutionary rate among different lineages.
```

DAMBE computes, using codes in the BASEML program in the PAML package, the log-likelihood for the tree without assuming a molecular clock (lnL = -4441.05) and that assuming a molecular clock (lnL = -4442.89). The likelihood ratio chi-square is calculated as $2*(\ln L_{no\ clock} - \ln L_{clock})$, which follows a chi-square distribution with n -1 degrees of freedom, where n is the number of species. Note that the likelihood ratio chi-square is the same as previously defined, i.e., $-2\ln(\lambda)$.

Based on the sample output, we clearly cannot reject the null hypothesis. In other words, the COI clock does not tick at significantly different rates in different hominoid lineages. This may also be true among different human lineages. This lends some indirect empirical support for the search for the mitochondrial Eve.

Chapter 21

Testing Phylogenetic Hypotheses

In previous chapters we have learned the fundamentals of phylogenetic reconstruction. We reconstructed phylogenetic trees without evaluating their statistical support relative to alternative trees. In this chapter we will learn the fundamentals of hypothesis testing in phylogenetics.

The true phylogeny can only be approximated. In phylogenetics, we often have quite a number of approximations represented by alternative phylogenetic hypotheses. One major task for phylogeneticists is to evaluate statistical support for each of the alternative hypotheses, with the hope that the majority of the alternative hypotheses can be eliminated and, in the ideal situation, only one will come out unrejected. This remaining hypothesis can then be used as a working hypothesis for other phylogeny-based studies, such as those using the comparative methods (Felsenstein 1985a; Felsenstein 1988a; Harvey and Keymer 1991; Harvey and Pagel 1991; Harvey and Purvis 1991).

Three subjects will be covered in this chapter. We first review some basic statistical concepts related to significance tests, especially those involving multiple comparisons. These concepts include Type I and Type II errors and the comparisonwise and experimentwise error rates. We then proceed to learn how to evaluate relative support for alternative phylogenetic relationships and the rationale behind such evaluation. This evaluation can be carried out by using either the distance method, the maximum parsimony method, or the maximum likelihood method. Finally, resampling techniques such as bootstrapping and jackknifing will be briefly discussed.

1. BASIC STATISTICAL CONCEPTS

A statistical significance test is always associated with a null hypothesis. A beginning student often think that a null hypothesis should be something absurd, proposed only to be rejected. This is not true. A null hypothesis that is absurd should never be formulated because the rejection of such a null hypothesis does not improve our understanding of nature. A good null hypothesis should always represent the conventional belief prior to our work, so that the rejection of the null hypothesis leads to a revision of our existing knowledge. We formulate a null hypothesis in science only when we have sufficient evidence to challenge the conventional wisdom.

The Type I error in a significance test is the error of rejecting a correct null hypothesis, and the Type I error rate the probability of making a Type I error and is typically represented by the Greek letter α. Type II error is the error of accepting a false null hypothesis, and the error rate is typically represented by the Greek letter β. Type II error is also called the consumer's error. For example, suppose a manufacturer puts only one kilogram of sugar into a package labelled as two kilogram, and suppose that everyone believes that the label on a package is a reliable indicator of the content inside. If we accept this conventional belief and pay the price of two kilograms of sugar for this one-kilogram package, then we, as consumers, are committing a Type II error. One can avoid making Type II errors by not accepting any hypothesis when the null hypothesis is not rejected. The two types of error is often summarized in the following table:

		Null Hypothesis	
		Accepted	Rejected
Null	TRUE	Correct Decision	Type I Error
Hypothesis	FALSE	Type II Error	Correct Decision

When we carry out a significance test, we typically would report a p value such as $t = 3$, $DF = 20$, $p = 0.0071$. This p value tells us is the Type I error rate. In other words, it is the probability that we would be wrong if we reject the null hypothesis. What does it mean if the p value is nearly 1? Surely it means that we would be almost 100% wrong if we reject the null hypothesis. But does not mean that we would be almost 100% correct if we accept the null hypothesis?

The answer is no. A p value of 1 only means that the evidence we have is perfectly consistent with the null hypothesis, but the evidence does not prove that the null hypothesis is true. For example, suppose a psychologist wanted to know if there is any difference in IQ score between adult males and females. He measured IQ for four males and three females and found the mean for the males to be identical to that for the females. If he used the t-test to test the null hypothesis of equal IQ between males and females, then the t statistic would be zero and the p value be 1. The data set is perfectly compatible with the null hypothesis, but its support for the null hypothesis is actually quite weak given the small sample size. So the psychologist is not guaranteed to be 100% correct if he accepts the null hypothesis as true.

This book only requires the reader to know when to reject a null hypothesis, not when to accept a null hypothesis. Consequently, we do not need to be concerned with the Type II error rate.

When a statistical test involves multiple comparisons, two confusing concepts arise: the Type I comparisonwise error rate (α_c) and the Type I experimentwise error rate (α_e). The former refers to the probability of making a Type I error in a single experiment, and the latter refers to the probability of making at least one type I error in N experiments. Let me offer an illustrative example where the failure to control for experimentwise error rate would lead to wrong conclusions.

Suppose we take two samples from the same normally distributed population, and test the difference in means between the two sample. As the two samples are taken from the same population, we expect them to have the same mean. However, the two means will typically not be identical because of sampling error, especially when sample size is small. If we repeat the sampling infinitely, then eventually we will obtain two samples with means significantly different from each other by a conventional t-test. If we then conclude that the two samples are from two different populations with different means, then we would be drawing a wrong conclusion - we know a priori that the two samples are from the same population.

If $\alpha_c = 0.05$, and N hypotheses are tested, then $\alpha_e \approx 1 - (1 - \alpha_c)^N$. If we have five trees, then there are a total of 10 pair-wise comparisons between the trees. Thus, $\alpha_c = 0.05$ would imply $\alpha_e \approx 0.40$. That is, if all trees are in fact equally good, there is still a probability of 0.4 that at least one tree will be incorrectly rejected. If N is infinite, then $\alpha_e = 1$, i.e., we are doomed to make an experimentwise error if α_c is not infinitely small.

If we are to control the experimentwise error rate below 0.05, we can set $\alpha_e = 0.05$, so that

$$\alpha_e \approx 1 - (1 - \alpha_c)^N = 1 - (1 - \alpha_c)^{10} = 0.05 \tag{21.1}$$

and solve the equation, which yields $\alpha_c = 0.005$. This of course would increase the difficulty to reject a null hypothesis, even if the null hypothesis is false.

The Student-Newman-Keuls test is appropriate for significance test involving multiple comparisons. It is used in DAMBE for all tests of alternative phylogenetic hypotheses implemented in DAMBE, including the distance, maximum parsimony, and maximum likelihood methods. The test statistic, q, is based on the same statistic as in Tukey's HSD test (Zar 1996). The rationale of the significance tests is outlined in the following sections dealing with individual tests using the distance, maximum parsimony and maximum likelihood methods.

2. TESTING PHYLOGENETIC HYPOTHESES WITH DISTANCE METHODS

2.1 The Rationale

Both neighbor-joining and Fitch-Margoliash method can be used for testing alternative phylogenetic hypotheses. The method for testing phylogenetic hypotheses by using the neighboring-joining method together with the minimum evolution (ME) criterion has already been developed and implemented in a computer program (Rzhetsky and Nei 1994). Three steps are involved. First, one makes a heuristic search of all topologies that are close to a ME tree. Second, branch lengths are estimated by the least-squares method. Third, the variance of branch lengths is estimated by resampling methods such as bootstrapping or jackknifing. Fourth, comparisons are made between the tree with the shortest branch lengths and alternative trees to see if the former is significantly shorter than all alternative trees.

The test implemented in DAMBE does not use the neighbor-joining method, but instead uses the Fitch-Margoliash method. The test is based on the fit to the original distance matrix. DAMBE reads in each of the **unrooted** user trees in a tree file. These user trees represent alternative phylogenetic hypotheses to be evaluated. Based on any one of many genetic distances you choose, DAMBE will calculates the distance matrix and evaluate tree branch lengths for each of the unrooted topologies in the user tree file, by using the least-squares method. From the reconstructed phylogenetic tree with optimized branch lengths, DAMBE then obtains the pairwise distances based on the estimated tree branch lengths. We thus have two matrices, one being the original distance matrix used for phylogenetic reconstruction, and the other being the reconstituted matrix derived from the reconstructed tree.

Let us designate the original matrix as **x** and the reconstituted matrix as **y**. Every element in **x** has one corresponding element in **y**. Elements in the **x** matrix represent observed values, and those in the **y** matrix represent the estimated or expected values. If $x_{ij} = y_{ij}$ for all i and j, then the fit between the observed and the estimated values is perfect, and the tree is considered to be the best possible.

DAMBE calculates the error variance as

$$VarE = \frac{\sum_{i=1}^{n-1}\sum_{j=i+1}^{n}(x_{ij}-y_{ij})^2}{n(n-1)/2-1} \tag{21.2}$$

where n is the number of species. No weighting by branch lengths is used. VarE is equivalent to the variance of D_{ij} (= x_{ij} - y_{ij}) because the mean D is expected to be zero. In other words, (D_{ij} - $\overline{D}$) = (Dij - 0) = (x_{ij} - y_{ij}).

If VarE = 0, then the tree is the best possible. For k alternative phylogenetic hypotheses, there are k VarE values. If these VarE values are all equal, then all trees are equally supported. If these VarE values are significantly different, then at least one of the trees is worse than the best-fit tree.

Testing whether all alternative topologies are equally good is now reduced to a test of heterogeneity of variance for which we have the standard Bartlett's test (Zar 1996, p. 204 and the references therein). The test is based on B_c = B/C where

$$B = (\ln s_p^2)\left(\sum_{i-1}^{k} v_i\right) - \sum_{i=1}^{k} v_i \ln s_i^2 = (N-2)\left[(\ln s_p^2)k - \sum_{i=1}^{k}\ln s_i^2\right] \tag{21.3}$$

$$\begin{aligned} C &= 1 + \frac{1}{3(k-1)}\left(\sum_{i=1}^{k}\frac{1}{v_i} - \frac{1}{\sum_{i=1}^{k} v_i}\right) \\ &= 1 + \frac{1}{3(k-1)}\left(\frac{k}{(N-2)} - \frac{1}{k(N-2)}\right) \end{aligned} \tag{21.4}$$

$$s_p^2 = \frac{\sum_{i=1}^{k} SS_i}{\sum_{i=1}^{k} v_i} = \frac{\sum_{i=1}^{k} SS_i}{k(N-2)} \tag{21.5}$$

In our case, the number of degree of freedom, v_i, is always the same and equals N – 2, where N is the number of D_{ij} values in either a lower or an upper triangular matrix. The reason for the degree of freedom to be equal to N – 2 rather than N – 1 is because one of the estimated y_{ij} values will always be the same as the corresponding x_{ij} value, and therefore will not contribute to the difference in variance. The distribution of B_c is approximated by the chi-square distribution with k-1 degrees of freedom (Nagasenker 1984), where k is the number of phylogenetic hypotheses.

The Bartlett's test is an overall test of homogeneity of variance. If the null hypothesis of equal variance is not rejected, then one should not proceed further. However, if the null hypothesis is rejected, i.e., some trees are significantly better (having smaller variances) than others, then one naturally would like to know if the best tree is better than all the others. It is not appropriate to do an F-test between VarE of the best tree and each of the (k-1) $VarE_i$ values because this does not control for experimentwise error rate.

To adjust for the experimentwise error rate associated with multiple comparisons, a test analogous to Newman-Keuls test is applied with the test statistic q (Zar 1996) calculated as

$$q = (\ln(VarA)-\ln(VarB))/SE \tag{21.6}$$

The main advantage of the test is that it should be very powerful in rejecting poor topologies. The quality of the test, of course, depends much on how appropriate the genetic distances are. The main weakness of the test is that the distances in the distance matrix are not statistically independent of each other, with the consequence that the test is overly conservative. Data from a limited simulation study (unpublished) suggest that this test, with an obviously wrong assumption, yields results very similar to the maximum likelihood method introduced later in the chapter.

2.2 Test alternative phylogenetic hypotheses with the distance method by using DAMBE

Suppose we are interested in the phylogenetic relationships among the human, the chimpanzee and the gorilla. There are only three possible phylogenetic groupings, i.e., human-chimpanzee, human-gorilla, and chimpanzee-gorilla. We will use the mitochondrial cytochrome oxidase subunit I (COI) gene for testing the three alternative phylogenetic hypotheses.

Start DAMBE and open a sequence file, e.g., the **Ape5.fas** file that comes with DAMBE. Click **Phylogenetics|Distance methods|nucleotide sequences** (You could choose a codon-based distance if your input sequences code for a protein gene, as is the case for the **Ape5.fas** file). A dialog box appears for you to choose options (fig. 1). The **User tree** appears only when the Fitch-Margoliash option is chosen because the neighbor-joining method does not take user trees. The user trees can either come from a file containing unrooted trees in PHYLIP format or, when the number of

OTU's is small, from all possible trees generated by DAMBE (fig. 1). We will use the following three alternative trees saved in a file named **Ape5ur.nhm** (where ur stands for **unrooted**):

```
(((CHBCO1,GGMCO1),HSMCO1),ORACO1,HLMCO1);
(((CHBCO1,HSMCO1),GGMCO1),ORACO1,HLMCO1);
(((HSMCO1,GGMCO1),CHBCO1),ORACO1,HLMCO1);
```

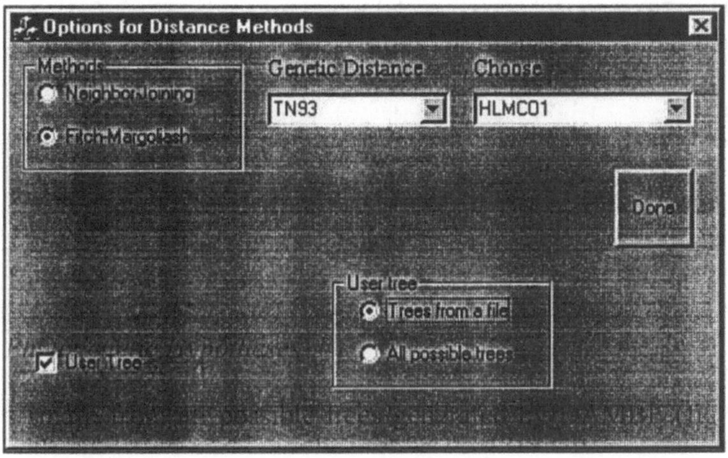

Figure 1. Evaluate alternative phylogenetic hypotheses with the Fitch-Margoliash method

When you click the **Trees from a file** option, a standard **File/Open** dialog box will appear for you to choose the tree file containing alternative trees to be evaluated. Click the **Ape5ur.nhm** file or any other relevant tree file containing unrooted trees. It is important that the name of the OTU's in the tree file is the same as the name of OTU's in the sequence data file that we have just read into DAMBE.

Click the **Done** button, and DAMBE will evaluate the branch lengths of alternative phylogenetic hypotheses one by one. You will be asked to enter the name of the file containing the re-evaluated trees, and of a file for saving output. Part of the output is reproduced below:

```
==============================================================
Performing Bartlett's test of equal variance with correction:

Bartlett's Bc =   1.511
DF =  2
Prob. =   0.4698
```

```
The null hypothesis of equal variance is not rejected. One
should not interpret any significant results shown below.
==============================================================
The tests in Part I have not adjusted for experimentwise error
rate, and are therefore not appropriate for multiple
comparisons. However, such tests have been done frequently in
literature.

The tests in Part II are based on Student-Newman-Keuls test
for multiple comparisons, and have been adjusted for
experimentwise error rate.

Part I:
==============================================================
Tree   Tree Length      Variance       F           Prob
 1       0.36488        0.00216      2.264       0.1345
 2       0.36222        0.00096  <===Best-fit tree
 3       0.36488        0.00216      2.264       0.1345
==============================================================

Part II:
==============================================================
Performing Student-Newman-Keuls test for multiple comparisons,
assuming large-sample variances.

The q(i) is equivalent to the t statistic and, if you wish,
might be declared significant at 0.05 level if q(i) > 1.96.
Such interpretation, however, is valid only when there are
only two topologies being compared.

Prob(i)-the probability of tree i being as good as
        the best tree, i.e., tree 2
k(i)   -the number of variances falling with the range
        between Var(i) and MinVar, including
        Var(i) and MinVar.
==============================================================
Tree       Var(i)        q(i)    k  Prob(i)
   1      0.00216      1.6341    2   0.2481
   2      0.00096 <====Best tree.
   3      0.00216      1.6341    3   0.4789
```

The Bartlett's test shows whether anyone of the three alternative trees is significantly better or worse than any other trees. There are only two degrees of freedom because we have only three trees. The result shows that the three trees are not significantly different from each other, i.e., they fit the observed data roughly equally well.

The **Prob.** value in the output is the probability that we would be wrong if we reject the null hypothesis. Here the null hypothesis is simply that the three $VarE_i$ values are the same. Our data set is insufficient to reject the null hypothesis.

The Bartlett's test is an overall test, and we need not proceed further if the test result is not significant, which is true in our case. However, if Bartlett's test is significant, then we would like to know which tree is significantly better (or worse) than others and, in particular, whether the best tree is significantly better than all the other alternatives. This test is done in two parts. Part I is simply a conventional F-test which do not control for experimentwise error rate. Part II presents the Newman-Keuls test for multiple comparisons, which does control for experimentwise error rate.

3. TESTING PHYLOGENETIC HYPOTHESES WITH THE MAXIMUM PARSIMONY METHOD

For the maximum parsimony method, a ROOTED tree is required to represent alternative topologies. This limitation arises from DAMBE's inheritance of the code from DNAPARS in PHYLIP. DNAPARS has already provided a significance test if you include user trees in the **infile**. In short, DNAPARS computes the number of steps (changes in character states) for each topology, the difference in the number of steps between the best and each alternative topology, and the associated (large-sample) variance of the differences. The z-score is computed and declared as significant if it is larger than 1.96 (Felsenstein 1985a). The main problem with this test is that the result can be interpreted probabilistically only when you have just two topologies and is not appropriate with multiple comparisons. DAMBE uses Newman-Keuls test that is valid for multiple comparisons.

Start DAMBE and open a sequence file, e.g., the **Ape5.fas** file that comes with DAMBE. Click **Phylogenetics|DNAMP |Nucleotide sequences**. A dialog box appears for you to choose options (fig. 2). Check the **User Tree**, and the two **User Tree** option buttons appear. Click the **Trees from a file** option. A standard **File/Open** dialog box appears for you to choose the tree file containing rooted trees to be evaluated. We will use the following three alternative trees saved in a file named **Ape5.nhm**:

```
((((CHBCO1,GGMCO1),HSMCO1),ORACO1),HLMCO1);
((((CHBCO1,HSMCO1),GGMCO1),ORACO1),HLMCO1);
((((HSMCO1,GGMCO1),CHBCO1),ORACO1),HLMCO1);
```

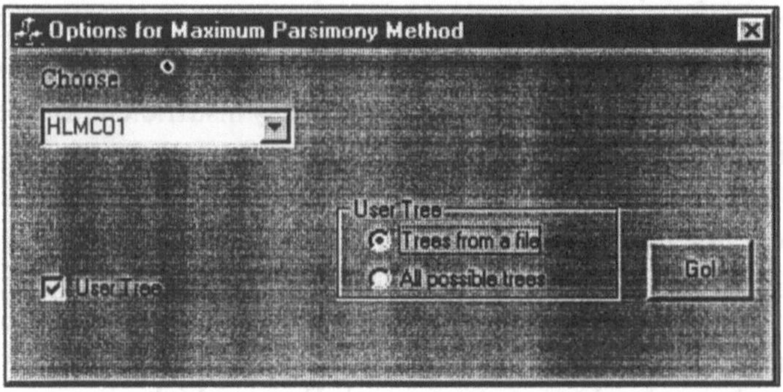

Figure 2. Evaluate alternative phylogenetic hypotheses with the Fitch-Margoliash method

Click the **Ape5.nhm** file or any other relevant tree file containing rooted trees. It is important that the name of the OTU's in the tree file is the same as the name of OTU's in the sequence data file that we have just read into DAMBE.

Click the **Go!** button, and DAMBE will evaluate the number of steps and its associated variance for each of the alternative phylogenetic hypotheses one by one. You will be asked to enter the name of the file containing the re-evaluated trees, and of a file for saving output. Part of the output is reproduced below:

```
Phylogenetic analysis using the maximum-parsimony method
implemented in Joe's DNAPARS, with minor modifications.

Evaluating 3 user trees.
Outgroup: HLMCO1

3 ROOTED trees evaluated:
((((CHBCO1,GGMCO1),HSMCO1),ORACO1),HLMCO1);
((((CHBCO1,HSMCO1),GGMCO1),ORACO1),HLMCO1);
((((HSMCO1,GGMCO1),CHBCO1),ORACO1),HLMCO1);

================================================================
Performing Student-Newman-Keuls test for multiple comparisons,
assuming large-sample variances.

The conventional test, as implemented in DNAPARS in PHYLIP,
ignores k and declares significant at 0.05 level if q(i) >
1.96.
```

```
Such interpretation, however, is valid only when there are
only two topologies being compared.

Prob(i)-the probability of tree i being as good as
        the best tree, i.e., tree 1
k(i)   -the number of means falling within the range
        between Step(i) and MinStep, including
        Step(i) and MinStep.
=====================================================================
Tree      Step(i)  Step(i)-Min  Std Dev(i)      q(i)    k  Prob(i)
   1     548.0000 <====Best tree.
   2     550.0000       2.0000      5.8328  0.3429    2    > 0.5
   3     553.0000       5.0000      5.5696  0.8977    3    > 0.5
```

Although the first tree requires the smallest number of changes, the difference is not significant between this tree and the other alternative trees. In this sense the result is consistent with the previous result from the distance method, i.e., the three trees describe data roughly equally well. However, you should remember that the test involving the distance method uses unrooted trees, and the test involving the maximum parsimony method uses rooted trees. Thus, the results are not really comparable between the distance method and the maximum parsimony method.

4. TESTING PHYLOGENETIC HYPOTHESES WITH THE MAXIMUM LIKELIHOOD METHODS

For the maximum likelihood method, the Kishino-Hasegawa test (Kishino and Hasegawa 1989) which is also called the RELL test, is implemented as in PAML (Yang 1997b) from which I have taken part of the code, again with the adjustment for multiple comparisons. The Kishino-Hasegawa test, as is practised in literature, is analogous to the test in DNAPARS, except that the test is based on the likelihood values rather than on the number of steps. In short, you calculate the log-likelihood for each topology, the difference in log-likelihood between the best tree and each of the alternative topologies, and the variance of the differences estimated by resampling methods such as bootstrapping. The z-score is then calculated and declared as significant if it is larger than 1.96. Again, such interpretation is heuristic and is not appropriate probabilistically if there are more than two topologies being compared. DAMBE uses Newman-Keuls test that is more appropriate for multiple comparisons.

To perform the test in DAMBE, you open a sequence file with aligned sequences, such as the **ape5.fas** file that comes with DAMBE. Click **Phylogenetics|Maximum likelihood|Nucleotide sequences**. When the dialog box (fig. 3) appears, click the **User tree** option. A standard **File/Open** dialog box appears for you to choose the file containing user trees in the PHYLIP format (alternatively you may click the option for all possible trees if the number of OTUs is small, e.g., 5 or 6). Let us again use the following three alternative trees saved in a file named **Ape5ur.nhm** (the same tree file used with the distance method):

```
(((CHBCO1,GGMCO1),HSMCO1),ORACO1,HLMCO1);
(((CHBCO1,HSMCO1),GGMCO1),ORACO1,HLMCO1);
(((HSMCO1,GGMCO1),CHBCO1),ORACO1,HLMCO1);
```

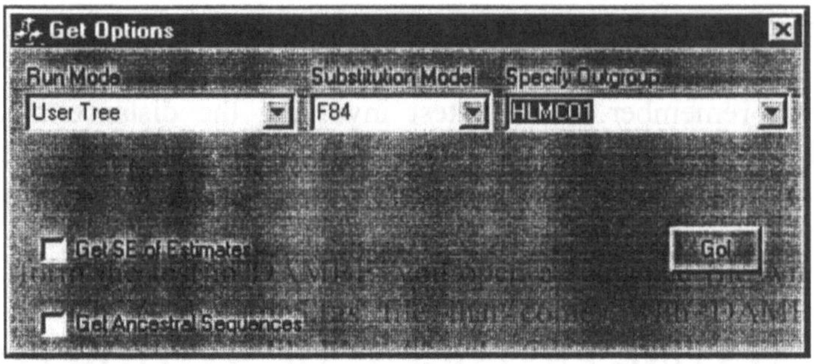

Figure 3. Testing alternative phylogenetic hypotheses by using the maximum likelihood method

Note that as soon as you have opened the tree file, the **clock** checkbox disappears. DAMBE will decide whether to impose a molecular clock or not by checking the first tree in the tree file. If the tree is rooted, i.e., the deepest node is bifurcating, then the tree is taken as rooted and a molecular clock is imposed. If the deepest node is multifurcating, then the tree is taken as unrooted and no clock is imposed. According to this criterion, the trees in the file **Ape5ur.nhm** are taken as unrooted and no clock is imposed.

Note that imposing a molecular clock may change the log-likelihood value substantially. Tree 1 could have a larger log-likelihood value than tree 2 when no molecular clock is imposed, but the reverse may be true when a molecular clock is imposed. It is important to have alternative phylogenetic hypotheses expressed either as all rooted trees or all unrooted trees, but not as a mixture of both because it is meaningless to compare relative statistical support for a rooted topology with that for an unrooted topology.

Also note that the test involving maximum likelihood method can take a tree file with either all rooted or all unrooted trees. We will use the unrooted trees in the **Ape5ur.nhm** first. The tree file **Ape5.nhm** contains all rooted trees and will be used later for evaluating alternative hypotheses.

Click the **Go!** button and DAMBE will compute the log-likelihood for each topology, the difference in log-likelihood between the best tree and all alternative trees, the variance of the difference by bootstrapping, and then perform the Student-Newman-Keuls test (It may take a long time with the maximum likelihood method). Part of the output is shown below:

```
Performing Student-Newman-Keuls test for multiple comparisons,
assuming large-sample variance.

The conventional test, as often practised in literature,
ignores k and declares significant at 0.05 level if q(i) >
1.96. Such interpretation, however, is valid only when there
are only two topologies being compared.

Prob(i)-the probability of tree i being as good as
        the best tree, i.e., tree 2
k(i)    -the number of ML values falling with the range
        between ML(i) and MaxML, including ML(i) and MaxML.
==============================================================
Tree     ML(i) ML(i)-MaxML SE_Dif(i) RELL(i)   q(i)   k  Prob(i)
 1    -4449.78     -8.7362   10.1485 0.1862  0.8608  2     > 0.5
 2    -4441.05 <====Best tree.
 3    -4455.55    -14.5055    8.9180 0.0150  1.6265  3    0.4823
```

Note that the result is similar to that derived from the test using the Fitch-Margoliash method, with the second tree being the best. However, there is no significant difference among the three trees. Note that the trees tested with the distance and the maximum likelihood methods are **unrooted**, and the results are similar. As the distance method is much faster than the likelihood method, and the number of sequences used in phylogenetic analysis is increasing rapidly, I expect the distance method to be used more frequently than it is today.

We have learned how to evaluate alternative phylogenetic hypotheses expressed in alternative unrooted trees. The maximum likelihood method can also be used to evaluate phylogenetic hypotheses expressed in rooted trees, which is what we are going to do now.

Click **Phylogenetics|Maximum likelihood|Nucleotide sequences**. When the dialog box (fig. 3) appears, click the **User tree** option. A standard **File/Open** dialog box appears for you to choose the file containing user trees in the PHYLIP format (alternatively you may click the option for all possible trees if the number of OTUs is small, e.g., 5 or 6). Click the **Ape5.nhm** (the same tree file as is used with the maximum parsimony method).

Click the **Go!** button and DAMBE will compute the log-likelihood for each topology, the difference in log-likelihood between the best tree and all alternative trees, the variance of the difference by bootstrapping, and then perform the Student-Newman-Keuls test. Part of the output is shown below:

```
Performing Student-Newman-Keuls test for multiple comparisons,
assuming large-sample variance.

The conventional test, as often practised in literature,
ignores k and declares significant at 0.05 level if
q(i) > 1.96. Such interpretation, however, is valid only when
there are only two topologies being compared.

Prob(i)-the probability of tree i being as good as
        the best tree, i.e., tree 1
k(i)   -the number of ML values falling with the range
        between ML(i) and MaxML, including ML(i) and MaxML.
==============================================================
Tree    ML(i) ML(i)-MaxML SE_Dif(i) RELL(i)    q(i)  k  Prob(i)
  1 -4442.89 <====Best tree.
  2 -4456.50    -13.6147    9.8885  0.0754  1.3768  2   0.3293
  3 -4461.417   -18.5320    8.7478  0.0022  2.1185  3   0.2950
```

By comparing the output with that from the maximum likelihood method, we notice that the maximum likelihood method seems to be more sensitive in detecting the difference among trees. than the maximum parsimony method, with Prob(i) being smaller in the latter than in the former. The difference, however, is negligible. The results are comparable because both are based on the same set of rooted trees.

5. RESAMPLING METHODS

A tree (i.e., a phylogenetic hypothesis) is made of one or more subtrees (i.e., subhypotheses). It is of interest to know the relative statistical support of alternative subtrees. Resampling methods in phylogenetics are for attaching confidence limits to these subtrees (Felsenstein 1985b).

There are two kinds of commonly used resampling methods (Efron 1982), one being bootstrapping and the other being delete-half jackknifing. The bootstrap resampling involves re-sampling the sample with replacement, so that sequences of N sites long will be resampled N times to generate new sequences of the same length, with some sites in the original sequences sampled one or more times while some other sites do not get sampled. The delete-half jackknifing technique will randomly purge off half of the sites from the original sequences so that the new sequences will be half as long as the original. Such resampling procedure will typically be repeated a large number of times to generate a large number of new samples.

Each new sample (i.e., new set of sequences), no matter whether it is from bootstrapping or jackknifing, will then be subject to regular phylogenetic reconstruction. The frequencies of subtrees will then be counted from reconstructed trees. If a subtree appears in all reconstructed trees, then the bootstrapping or jackknifing value is 100%, i.e., the strongest possible support for the subtree.

Although the bootstrap and the jackknife generally produce similar results, there are some subtle differences. Suppose that we have a set of aligned sequences of N sites long. For the bootstrap resampling, the probability of each site being sampled is 1/N, and the mean number of times a site gets sampled in each bootstrapping resampling is simply one. Thus, a site gets sampled 0, 1, 2, ..., N times follows a Poisson distribution with mean equal to one. This implies that about 37% of the sites will not be sampled, while 63% of the sites will be sampled at least once. In jackknifing, we have 50% of the sites not sampled and the other 50% of the sites sampled just once. Thus, a jackknived sample is expected to be less similar to the original sample than a bootstrapped sample. Consequently, jackknived samples should be less similar to each other than bootstrapped samples.

Both the bootstrap and the jackknife resampling methods are implemented in DAMBE in conjunction with phylogenetic analysis using the distance and the maximum parsimony methods. To see how it works, just start DAMBE and open an aligned sequence file, e.g., the **invert.fas** file that comes with DAMBE. Click **Phylogenetics|Distance methods|Nucleotide sequences**. A dialog box appears for you to specify options. Click the **Resampling statistics** checkbox. Two option buttons will appear, one labelled **Bootstrap** and the other **Jackknife**, with the default being the bootstrap. Below the option buttons is a text-input field for you to specify the number of resampled data sets. The default value is 100. Click the **Done** button, and DAMBE will resample the data and produce a consensus tree showing resampling support for the subtrees.

The implementation and the use of the resampling methods in conjunction with the maximum parsimony method is essentially the same as above. Just click **Phylogenetics|DNAMP|Nucleotide sequences** and follow what you have done with the distance method.

6. EXERCISE

Find a paper on phylogenetic analysis involving DNA sequences. Derive from the paper alternative phylogenetic hypotheses and express these hypotheses in tree topologies in PHYLIP format. Evaluate these alternative phylogenetic hypotheses by using various distance, maximum parsimony and maximum likelihood methods. Write a report to elaborate your results in reference to the conclusions drawn by the author(s) of the paper.

Chapter 22

Fitting Probability Distributions To Substitutions Over Sites

1. INTRODUCTION

This chapter deals with fitting probability distributions to the pattern of nucleotide, amino acid and codon substitutions along the molecular sequences. It is important to know if the substitution rates vary among sites, because such rate heterogeneity, according to a comparative study based on simulated data (Kuhner and Felsenstein 1994), would results in failure to recover the true phylogenetic relationships in virtually all commonly used phylogenetic programs (or algorithms), including maximum likelihood method (e.g., PHYLIP), maximum parsimony (e.g., PAUP), or neighbor-joining (e.g., MEGA) methods.

Fitting probability distributions to substitution data requires a sequence file with reconstructed ancestral sequences as well as a tree that has been used for reconstruction of ancestral sequences. Please read the last section of the chapter dealing with molecular phylogenetics on how to reconstruct ancestral sequences by using DAMBE. If you use the PAML package, then the output file for saving ancestral sequences is named **rst** by default. DAMBE can read this file directly if you rename the file to **SomeName.rst** **(SomeName** means any valid file name for the Windows platform).

Two discrete probability distributions, the Poisson and the negative binomial, have been used to fit the distribution of substitutions along the molecular sequences. A continuous distribution, the gamma distribution, has recently been used extensively in modelling the heterogeneity of substitution rates over sites. I will first provide some statistical background for these discrete and continuous distributions and then illustrate their use in analysing molecular data.

1.1 The Poisson distribution

Nucleotide or amino acid substitutions are stochastic events, and the Poisson process is one of the simplest stochastic processes. The Poisson process is often described in two ways, one by the discrete Poisson distribution and one by the continuous gamma distribution. In the first case, the random variable is the number of events (e.g., substitutions) occurring during a fixed length of time, and is therefore discrete. In the second case, the random variable is the length of time until the rth occurrence of an event, and is therefore continuous. We will first concentrate on the discrete Poisson distribution.

The Poisson distribution can be viewed a special case of the binomial distribution that we have already studied in the chapter entitled "A statistical digression". The binomial theorem states that if the probability of a success in a single trial is p, then the probabilities of 0, 1, 2, ... successes out of n trials are given by the successive terms from the expansion of $(p + q)^n$, where $q = 1 - p$:

$$q^n + npq^{n-1} + \frac{n(n-1)}{2}p^2q^{n-2} + ... + \binom{n}{r}p^rq^{n-r} + ... + p^n$$
$$= q^n\left[1 + \frac{np}{q} + \frac{n(n-1)}{2}\left(\frac{p}{q}\right)^2 + \frac{n(n-1)(n-2)}{3!}\left(\frac{p}{q}\right)^3 + ...\right] \tag{22.1}$$

Equation (22.1) looks complicated but is in fact a simple equation. Just replace n by 2 or 3 and you see how simple it is. The binomial distribution has the mean equal to n•p and the variance equal to n•p•q, i.e., the mean is greater than the variance.

Suppose we are observing the occurrence of stochastic events, such as nucleotide substitutions, over a time period t. We may divide the period t into n intervals and record the number of occurrences of the event. The event has a probability of p to occur within each of n fixed time interval. Obviously the number of events that could happen within the time interval could be 0, 1, 2, ...x, with the mean equal to n·p. This is beyond what a binomial distribution can handle because a binomial distribution deals with statistical experiments with only two dichotomous outcomes, represented as the success and the failure.

One way to proceed is to divide the time period t into an infinitely large number of intervals, i.e., n = ∞, so that each interval will be so short that the probability of having more than one occurrence of the event within each fixed time period is negligible. Now we can view each fixed time interval as an independent Bernoulli trial, and the stochastic process during the time period t as a Bernoulli process with n very large and p very small.

Given that p is very small and n very large in a binomial distribution, then the right-hand term of equation (22.1) can be written as:

$$
\begin{aligned}
&(1-p)^n\left[1+np+\frac{(np)^2}{2}+\frac{(np)^3}{3!}+\ldots\right] \\
&= e^{-np}\left[1+np+\frac{(np)^2}{2}+\frac{(np)^3}{3!}+\ldots\right]
\end{aligned}
\tag{22.2}
$$

Three approximations were used to convert the right hand side of equation (22.1) to that of equation (22.2). First, when p is very small, then $q \approx 1$. Second, if n is very large, then $n(n-1)(n-2)...(n-r) \approx n^{r+1}$. Third, when p is small and n larger, $(1 - p)^n \approx e^{-np}$. Notice that the summation inside the brackets happens to e^{np}. So the sum of all the terms in the equation above is one, which makes sense because the probabilities are supposed to sum up to one. The successive terms represent the same meaning as in the binomial distribution, i.e., they are probabilities that there are 0, 1, 2, ... successes out of n trials. The probability distribution specified by these successive terms is called the Poisson distribution defined by

$$
P(r\,|\,\lambda)=\frac{e^{-np}(np)^r}{r!}=\frac{e^{-\lambda}\lambda^r}{r!} \tag{22.3}
$$

The Poisson distribution has the mean equal to the variance. It has been used in ecology for modelling a random distribution of individuals in space. We use it to discover whether substitutions are in fact randomly distributed along the molecular sequence (which I use to refer to either the nucleotide or amino acid sequence).

Fitting the Poisson distribution to substitution data is done as follows. First, molecular sequences of length N are divided into N sample sites, or N/n sample sites if we use n neighbouring sites as a single sampling unit. Second, the number of substitutions in each sampling site (unit) is counted by pair-wise comparison between neighboring nodes along a phylogenetic tree. Third, the number of sampling sites (units) containing 0, 1, 2 ... substitutions are compared with the number expected according to the

Poisson distribution. The extent of match between the expected numbers and the observed numbers can be tested by a Chi-squared goodness-of-fit test. If the two sets of numbers match well, then we conclude that the substitution pattern does not deviate significantly from the Poisson process. In other words, we do not reject the null hypothesis that the same Poisson clock ticks in all sites.

The Poisson distribution, when applied to model substitutions along molecular sequences, assumes that the probability of multiple substitutions at the same nucleotide or amino acid site is negligible, and that the substitution at one site is independent of other sites. These assumptions often cannot be met by substitutions along molecular sequences. Multiple substitutions at the same site is a reality, and the effect of neighboring nucleotides on the substitution rate is well documented (Bulmer 1986; Morton and Clegg 1995). However, there is nothing wrong in using the Poisson distribution as a null model for the distribution of substitutions along molecular sequences.

1.2 The negative binomial distribution

There are two reasons that the distribution of substitutions along molecular sequences should deviate from the Poisson distribution. First, some sites are more important than others, and consequently would be subject to stronger purifying selection and evolve more slowly than others. Take a codon for example, a nucleotide mutation at the second codon position invariably results in a nonsynonymous substitution, whereas a nucleotide mutation at the third codon position is frequently silent. Consequently, the substitution at the third codon position occurs much more frequently than that at the second codon position. Secondly, a gene at the DNA level is typically a structured entity, with some segments functionally more important than others. Consequently, some segments will be more conservative than others. For example, most mitochondrial proteins are transmembrane proteins made of hydrophobic and hydrophilic domains associated with different nonsynonymous substitution rates (Irwin et al. 1991; Kyte and Doolittle 1982). Hydrophobic domains are typically embedded inside the membrane and not associated with reaction centres. Consequently, these domains have higher nonsynonymous substitution rates than hydrophilic domains that are often associated with reaction centres.

The recognition of hydrophobic or hydrophilic segments is aided by a polarity plot, which I have done (by using DAMBE) for the ATPase 6 from human mitochondrial genome (Fig. 1). The structural heterogeneity of the protein molecule is obvious in Fig. 1. The distribution of nonsynonymous substitutions along the DNA sequences also exhibits apparent discontinuity, especially at the second codon position (Xia 1998b) where long stretches of the DNA sequences harbour no nonsynonymous substitutions at all for all 35 pair-wise comparisons among 19 mammalian species (Xia 1998b). This rate heterogeneity would results in variable substitution rate among sites, which in turn would lead to substitution patterns deviating significantly from the Poisson distribution.

The pattern of substitutions clumped at certain segments but lacking in other segments of molecular sequences will generate a right-skewed distribution for the number of substitutions per site. Such a distribution can be fitted by the negative binomial distribution, which is right-skewed and has the variance greater than the mean. The distribution has been used in ecology to model the clumped distribution of individuals in space (e.g., Krebs 1999; Xia and Boonstra 1992).

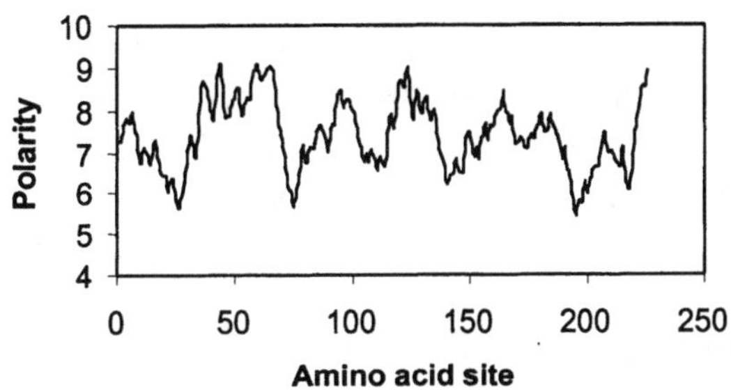

Figure 1. Polarity plot for human mitochondrial ATPase 6, generated by using DAMBE. Each point represents a moving average of polarity values for 10 neighboring amino acids. Polarity values are from Grantham (1974).

In contrast to the Poisson distribution which has just one parameter, the negative binomial distribution has two parameters, p and k, with the mean $\mu = kp$ and $\sigma^2 = \mu + \mu^2/k$. Both parameters affect the shape of the negative binomial distribution. When μ increases (i.e., when p and k increases), the negative binomial distribution will gradually converge to the normal distribution with $\mu = kp$ and $\sigma^2 = \mu + \mu^2/k$. When p is constant, then the parameter k, which is often called the negative binomial parameter, determines the shape of the distribution. The smaller the k, the greater the clumping. In terms of rate heterogeneity of substitutions over sites, a small k means a great rate heterogeneity and vice versa. In this aspect, k is similar to the shape parameter α in the gamma distribution where rate heterogeneity increases as α decreases. In general, the rate heterogeneity is greater among the second codon positions of protein-coding genes than that among the third codon positions, which you can easily verify by using DAMBE.

There is no analytical solution for the maximum likelihood estimator of k. DAMBE uses the maximum-likelihood estimator in Johnson et al (1992, p. 216) to estimate the negative binomial parameter k through computer iteration:

$$\hat{k}\,\hat{p} = \bar{x}$$

$$\ln(1-\hat{p}) = \sum_{j=1}^{\infty}\left(\frac{\sum_{i=j}^{\infty} f_j}{\hat{k}+j-1}\right) \tag{22.4}$$

where f_j is the observed number of sites experiencing j substitutions. The iteration stops when the difference between the two sides of the equation is smaller than 0.00001.

Procedures involved in fitting the negative binomial distribution to substitution data is similar to that for fitting the Poisson distribution, except that the two parameters characterizing the negative binomial distribution are estimated and used to compute the expected frequency distributions. Multiple substitutions at the same site is also assumed to be negligible.

1.3 The gamma distribution

One problem with fitting the discrete distributions to substitution data is as follows. Suppose you have made pair-wise sequence comparisons between neighboring nodes along a tree and found three substitutions at site 2, with two being transitions and one being transversions. What you have got is the minimum number of substitutions required to explain the molecular evolution at this site, but the sequences may well have experienced multiple hits at the same sites, with latter substitutions erasing previous ones. It is for this reason that many substitution models have been proposed to correct for multiple hits. If you apply one of these correction methods to correct for multiple hits, then you will get an estimated number of substitutions being larger than just three, and the estimated number will not be an integer. In other words, the number of substitutions at each site will be continuous rather than discrete. So discrete distributions such as the Poisson and the negative binomial will no longer be applicable.

The gamma distribution is a continuous distribution, and is similar to the discrete negative binomial distribution in that both are right-skewed. This makes it appropriate for modelling substitution patterns where some sites (e.g., functionally important DNA segments) vary little while some other sites (e.g., non-functional DNA segments) experience a lot of substitutions.

Many continuous random variables have a right-skewed distribution, e.g., the lengths of time between malfunction for aircraft engine, the lengths of time between arrivals at a supermarket checkout queue, the life span of electronic components such as fuses, or the lengths of time to complete a maintenance checkout for an automobile engine. These variables are frequently modelled by the gamma distribution.

The gamma distribution has recently been used extensively in phylogenetics. The distribution has two parameters, one being the shape parameter, often symbolized by the Greek letter α, and the other being the scale parameter, often symbolized by β:

$$p(x;\alpha,\beta)=\frac{x^{\alpha-1}e^{-x/\beta}}{\beta^{\alpha}\Gamma(\alpha)} \tag{22.5}$$

Let me explain why the parameter β is called the scale parameter. Suppose you have a random variable X that has the gamma distribution with shape parameter α and scale parameter β. It can be shown that the variable cX, where c is a constant larger than 0, also follows a gamma distribution, with the shape parameter α and scale parameter cβ. If we divide X by the scale parameter β, then the distribution of the resulting variable X/β, also follows a gamma distribution and is called a standard gamma distribution. Thus, if you have a computer function, say, **IncompGamma** (α, X_i) that calculates the cumulative distribution probability for the standard gamma distribution, then the cumulative distribution probability for the generalized gamma distribution is simply **IncompGamma** (α, X_i/β).

The shape of the gamma distribution is determined by the parameter α. Hence its name as the shape parameter. This represents one additional advantage of the gamma distribution over the negative binomial distribution where both parameters affect the shape of the distribution. When α is equal or smaller than 1, the mode of the distribution is at the smallest possible value, zero (fig. 2). When α is larger than 1, then the mode equals (α - 1)β. The gamma distribution converges to the normal distribution with mean αβ and variance $\alpha\beta^2$, when α approaches infinity (fig. 2). There are several probability distributions that are special cases of the gamma distribution. The chi-square distribution is a special gamma distribution with α = ν/2 and β = 2. The exponential distribution is also a special gamma distribution with α = 1. Now we have come to appreciate why the gamma distribution is a parametric family of distributions with a rich variety of shapes and why α is called a shape parameter.

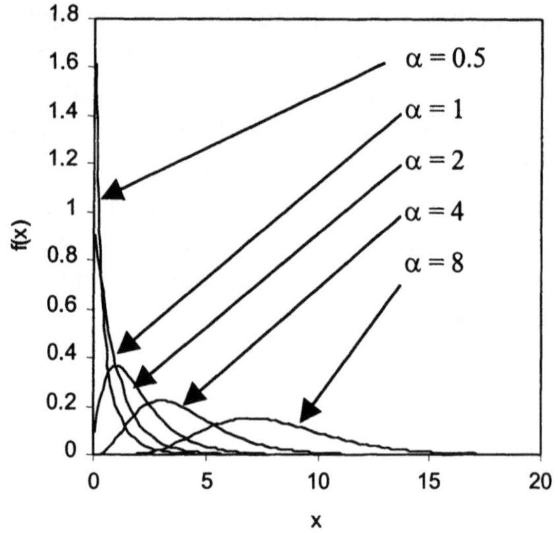

Figure 2. Gamma distributions with the shape parameter β = 1. Note that the shape of the distribution changes with the shape parameter α.

The α parameter can be used to measure the rate heterogeneity over sites. The smaller the α value, the greater the rate heterogeneity. Most sequences yield an α value near 0.5. For protein-coding genes, the α value is around 0.2 when estimated from second positions, but much larger when estimated from third codon positions.

It is important to remember that the observed substitution data is always discrete, and it is inherently awkward to fit a continuous distribution to the observed substitution data. First, the number of substitutions (designated x_i) for a given site cannot take a value larger than 0 but smaller than or equal to 1. For example, if the observed $x_i = 1$, then the estimated x_i, after correction for multiple hits, will take a value greater than one. If we observe $x_i = 0$, then its estimated value will be zero regardless of what substitution model you use. Second, when the number of sequences is small, then we will have a number of sites with the estimated x_i values slightly larger than 1, slightly larger than 2, and so on, with no site having values in between (fig. 3). Such a distribution is hardly continuous.

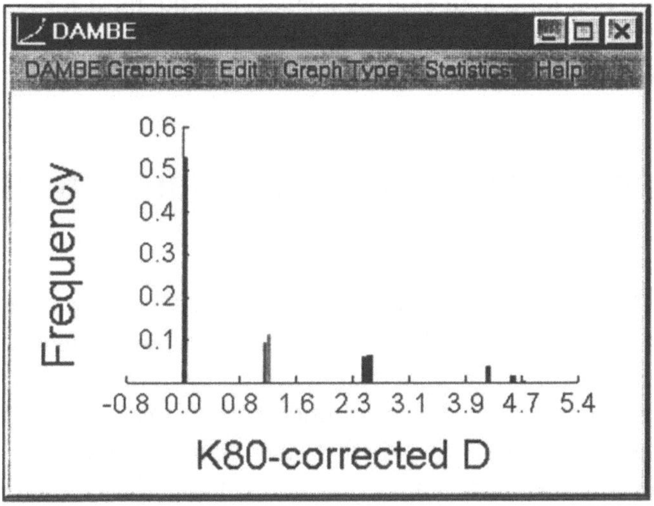

Figure 3. The frequency distribution of the number of substitutions per site, from five invertebrate species and the reconstructed ancestral sequences.

1.4 Some general guidelines for fitting statistical distributions

One should be aware that fitting statistical distributions to empirically derived substitution data requires the substitution data to be derived from pair-wise comparisons between neighboring nodes along a phylogenetic tree. It is simply meaningless to fit statistical distributions to substitution data derived from all possible pair-wise comparisons between input sequences. For example, if you have 8 sequences in your file, there are 28 possible pair-wise comparisons. When you look at the number of sites that have experienced 0, 1, 2, ..., N substitutions, you will find a number of sites with 0 substitutions, no sites experiencing 1, 2, 3, 4, 5, or 6 substitutions, yet a number of sites experiencing 7 substitutions (i.e., when one substitution occurred in one sequence but not in any of the other 7 sequences).

To do a proper fitting of statistical distribution to your sequences, you should make only independent comparisons. This can be done in two ways. First, if your input file has a tree topology (i.e., in RST format), then DAMBE will automatically take this tree topology and make comparisons only between neighboring nodes. This is the preferred method. Second, if your file is not in RST format, then you should choose sequence pairs in such a way that a sequence is compared with another sequence only once. Doing anything else is likely to lead to erroneous conclusions.

DAMBE provides three types of output when fitting statistical distributions, one for nucleotide-based sequences (i.e., sequences not coding for proteins), one for amino acid sequences, and the third for protein coding sequences. In the latter case, one can fit statistical distributions either to all substitutions or to nonsynonymous substitutions only.

2. FITTING DISCRETE DISTRIBUTIONS WITH DAMBE

DAMBE provides you with a few tools that you can use to test the distribution pattern of nucleotide or codon substitutions along the sequences. One can use DAMBE to first test whether the Poisson distribution can provide a reasonable fit to the substitution data. If the Poisson distribution is rejected, then one can test whether the negative binomial distribution provides an adequate fit. The fit to the gamma distribution is only for estimating the shape parameter α as a descriptive statistic of rate heterogeneity among sites. The shape parameter can also be used to correct the estimation of genetic distances. The gamma-corrected distances are implemented in DAMBE only with the F84 and TN93 models.

Start DAMBE and open a file. Preferably you should open a file in RST format (e.g., the **invert.rst** file that comes with DAMBE) so that all pair-wise comparisons of nucleotide, amino acid, or codon differences will be carried out between neighboring nodes along the tree topology. Once the sequences are displayed in DAMBE's display window, click **Seq. Analysis|Fit discrete distributions to substitutions**. A dialog box (fig. 4) appears for you to specify options. The first is to decide whether to use the Poisson distribution or the negative binomial distribution. The default is the Poisson distribution. The next option is the window size, i.e., you decide whether you should use a single site as a sampling unit, or whether you would use several neighboring sites as a single sampling unit. If the window size is 3, then 3 consecutive nucleotide or amino acid sites are used as a single sampling unit, and nucleotide sequences of 900 bases long would have 300 non-overlapping sampling units. The last option, i.e., **Nucleotide-based** or **Codon-based**, might be confusing to you. If your nucleotide sequences are from protein-coding genes and you have chosen the **codon-based** option, then DAMBE will present two more options, i.e., whether to fit the Poisson distribution only to nonsynonymous substitutions or to all substitutions over sites. The rate heterogeneity is usually greater for the nonsynonymous substitutions than for the synonymous substitutions (Xia 1998b). For the time being, we will do just a nucleotide-based analysis.

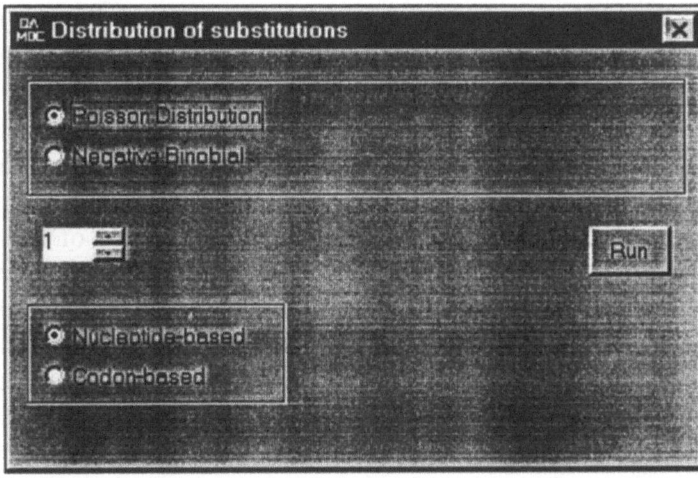

Figure 4. Fitting discrete distributions to substitution data over sites

Click the **Run** button, and DAMBE will prompt you for an output file for saving results. Enter a file name and the result will be saved in text format. Part of the result is shown below based on sequences in the **invert.rst** file that comes with DAMBE.

```
Fitting Poisson distribution to observed substitutions
along the nucleotide sequences.
Input file: D:\MS\DAMBE\invert.RST

Sequence Length:  1089  (after deleting gaps and unresolved
sites.)
=============================================
          observed   expected         Chi-
   x     frequency  frequency        square
---------------------------------------------
   0           576     394.78         83.19
   1           168     400.58        135.04
   2           156     203.23         10.98
   3           136      68.74         65.81
   4            49      17.44         57.13
>= 5             4       4.24          0.01
---------------------------------------------
Sum           1089    1089.00        352.16

P = 0.0000 with  4  degree(s) of freedom.
Lambda =  1.01. Variance =  1.65. Window size =  1
```

We see that 576 sites have not experienced any substitution. However, four sites have experienced five or more substitutions. The goodness-of-fit test shows that the Poisson distribution fits the observed substitution data poorly (P = 0.0000). Note that there are six groups, but only four degrees of freedom for the chi-square test. The reason for the loss of two degrees of freedom is as follows. First, we need the total number of sites to compute the expected values. Once the total is known, then only five of the observed frequencies are free to vary. For example, given that the total number of sampling sites is 1089, I need to have only five observed frequencies in order to deduce the sixth. Hence the loss of one degree of freedom. We also estimated the Poisson parameter, **lambda** (**λ**), which is also needed for computing the expected values. Hence the loss of another degree of freedom. So the degree of freedom associated with the chi-square test is down to four. Remember that a degree of freedom is lost whenever we estimate a parameter needed for carrying out the significance test.

The output is what we would have expected. The substitution rate of the three codon positions differs greatly from each other, with the third codon position the least conservative and the second codon position the most conservative (Xia 1998b; Yang 1996b). The result above calls for a goodness-of-fit test with the negative binomial distribution, which yields the following result:

```
Fitting negative binomial distribution to observed
substitutions along the nucleotide sequences.
Input file: D:\MS\DAMBE\invert.RST

Sequence Length: 1089 (after deleting gaps and unresolved
sites.)
=============================================
          observed   expected        Chi-
   x     frequency  frequency       square
---------------------------------------------
   0           576     540.66         2.31
   1           168     272.13        39.85
   2           156     137.06         2.62
   3           136      69.04        64.93
   4            49      34.78         5.81
   5             3      17.52        12.04
   6             1       8.83         6.94
   7             0       4.45         4.45
   8             0       2.24         2.24
>= 9             0       2.27         2.27
---------------------------------------------
Sum           1089    1088.99       143.46
```

```
P = 0.0000 with 7 degree(s) of freedom.
k = 0.9988
```

There are 10 frequencies, but only seven degrees of freedom associated with the chi-square test, i.e., we lost three degrees of freedom with the negative binomial distribution in contrast to the Poisson distribution where we lost only two degrees of freedom. This is because the negative binomial distribution is characterized by two parameters (k and p) and the Poisson distribution is a single-parameter distribution. For each parameter we estimate from the sample, we lose one degree of freedom. The fit is better than before, but still not satisfactory (P = 0.0000).

3. ESTIMATING THE SHAPE PARAMETER OF THE GAMMA DISTRIBUTION FROM SUBSTITUTION DATA

Start DAMBE and open a sequence file in RST format, e.g., the **invert.rst** file that comes with DAMBE. Once the sequences are displayed, click **Ana. Sequences|Fit gamma distribution to substitutions**. A standard **File/Open** dialog box will come up for you to enter a file name for saving results. You will also be asked whether you wish to have a graphic output of the empirical frequency distribution of the estimated number of substitutions per site. One such graphic output is shown in fig. 5.

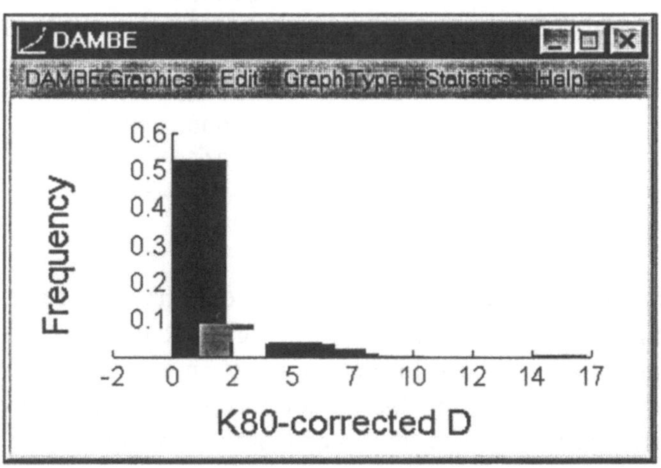

Figure 5. Empirical frequency distribution of the estimated nucleotide substitutions per site.

What DAMBE has done is as follows. First, the numbers of transitions and transversions, symbolized as N(s) and N(v), respectively, in the output, are counted for each site from pair-wise comparisons between neighboring nodes along the phylogenetic tree. Second, the estimated number of substitutions per site is then calculated by using Kimura's two-parameter method. These estimated numbers of substitutions per site are then used to obtain the shape parameter α of the gamma distribution. Part of the output is shown below:

```
Estimating the alpha parameter for the gamma
distribution, based on file:
D:\MS\DAMBE\inv7.RST

Computing the number of substitutions per site.
N(s)  - number of transitions.
N(v)  - number of transversions.
N(diff) - number of substitutions after correction
with Kimura's 2-parameter model.
================================================
Site          N(s)       N(v)        N(diff)
 1               0          0         0.0000
 2               0          0         0.0000
 3               1          0         1.1037
 4               0          0         0.0000
 5               0          0         0.0000
 6               1          0         1.1037
 7               0          0         0.0000
 8               0          0         0.0000
 9               2          0         2.4859
......
 1087            0          1         1.0761
 1088            0          0         0.0000
 1089            0          2         2.3466
================================================
The estimated alpha =   0.60137
```

The estimated α (= 0.6014) is what we would have expected by looking at the empirical frequency distribution in fig. 5. Recall that the mode of a gamma distribution will be at the zero point when α is equal or smaller than 1. The mode in the frequency distribution in fig. 5 is clearly zero, so that α should be smaller than 1. The frequency distribution (fig. 5) also shows that, while most sites are conservative, some sites have changed extremely fast, with one site having the estimated number of substitutions greater than 14.

4. EXERCISE

It has been shown for protein-coding genes that rate heterogeneity is the greatest at the second codon position, and the smallest at the third codon position (Xia 1998b; Yang 1996a). Use a set of protein-coding genes to estimate the alpha value for the first, second and third codon positions. Is the estimated alpha value the smallest at the second position and greatest at the third codon position? What could have caused this pattern?

Literature Cited

Adachi J, Hasegawa M (1996) Model of amino acid substitution in proteins encoded by mitochondrial DNA. Journal of Molecular Evolution **42**:459-468

Akashi H (1994a) Synonymous codon usage in Drosophila melanogaster: natural selection and translational accuracy. Genetics 136:927-935

Akashi H (1994b) Synonymous codon usage in Drosophila melanogaster: natural selection and translational accuracy. Genetics 136:927-35

Akashi H, Eyre Walker A (1998) Translational selection and molecular evolution. Curr-Opin-Genet-Dev 8:688-93

Altman PL, Dittmer DS (1972) Biology Data Book, Vol III 2nd ed. Federation of American Societies for Experimental Biology.

Antequera F, Bird A (1993) CpG Islands. In: Jost JP, Saluz HP (eds) DNA methylation: molecular biology and biological significance. Birkhäuser Verlag, Basel, p 169-185

Aquadro CF, Greenberg BD (1983) Human mitochondrial DNA variation and evolution: analysis of nucleotide sequences from seven individuals. Genetics 103:287-312

Archie JW (1989) A randomization test for phylogenetic information in systematic data. Systematic Zoology 38:219-252

Argos P, Rossmann MG, Grau UM, Zuber A, Franck G, Tratschin JD (1979) Thermal stability and protein structure. Biochemistry 18:5698-5703

Avise JC, Nelson WS, Sugita H (1994) A speciational history of 'living Fossils': molecular evolutionary patterns in horseshoe crabs. Evolution 48:1986-2001

Baldi P, Brunak S (1998) Bioinformatics. The MIT Press, Cambridge, Massachusetts

Barker D, Schafer M, White R (1984) Restriction sites containing CpG show a higher frequency of polymorphism in human DNA. Cell 36:131-138

Bechkenbach AT, Thomas WK, Homayoun S (1990) Intraspecific sequences variation in the mitochondrial genome of rainbow trout (Oncorhynchus mykiss). Genome 33:13-15

Bennetzen JL, Hall. BD (1982) Codon selection in yeast. J Biol Chem 257:3026-3031.

Bilgin N, Ehrenberg M, Kurland C (1988) Is translation inhibited by noncognate ternary complexes? FEBS.LETT. 233:95-99.

Boissonnas RA, Guttmann S (1960) Synthèse d'analogues de l'oxytocine et de la lysine-vasopressine contenant de la phénylalanine ou de la tyrosine en positions 2 et 3. Helvitica Chimica Acta 43:190-200

Bonekamp F, Andersen HD, Christensen T, Jensen KF (1985) Codon-defined ribosomal pausing in *Escherichia coli* detected by using the *pyrE* attenuator to probe the coupling between transcription and translation. Nucleic Acids Res. 13:4113-4123

Bonekamp F, Jensen KF (1988) The AGG codon is translated slowly in *E. coli* even at very low expression levels. Nucleic Acids Res. 16:3013

Bridger WA, Henderson JF (1983) *Cell ATP*. Wiley, New York.

Brown GG, Simpson MV (1982) Novel features of animal mtDNA evolution as shown by sequences of two rate cytochrome oxidase subunit II genes. Proc Natl Acad Sci USA 79:3246-3250

Brown WM, Prager EM, Wang A, Wilson A (1982) Mitochondrial DNA sequences of primates: tempo and mode of evolution. J Mol Evol 18:225-239

Bulmer M (1986) Neighboring base effects on substitution rates in pseudogenes. Mol. Biol. Evol. 3:322-329

Bulmer M (1988) Coevolution of codon usage and transfer RNA abundance. Nature **325**:728-730

Bulmer M (1991) The selection-mutation-drift theory of synonymous codon usage. Genetics 129

Cann RL, Stoneking M, Wilson AC (1987) Mitochondrial DNA and human evolution. Nature 325:31-36

Cavalli-Sforza LL, Edwards AWF (1967) Phylogenetic analysis: models and estimation procedures. Evolution 32:550-570

Clarke B (1970) Selective constraints on amino-acid substitutions during the evolution of proteins. Nature 228:159-160

Cooper DN, Krawczak M (1989) Cytosine methylation and fate of CpG dinucleotides in vertebrate genomes. Hum. Genet. 83:181-189

Cooper DN, Krawczak M (1990) The mutational spectrum of single base-pair substitutions causing human genetic disease patterns and predictions. Hum. Genet. 85:55-74

Cooper DN, Schmidtke J (1984) DNA restriction fragment length polymorphisms and heterozygosity in the human genome. Hum. Genet. 66:1-16

Cooper DN, Youssoufian H (1988) The CpG dinucleotide and human genetic disease. Hum. Genet. 78:151-155

Dayhoff MO, Schwartz RM, Orcutt BC (1978) A model of evolutionary change in proteins. In: Dayhoff MO (ed) Atlas of Protein Sequence and Structure. National Biomedical Research Foundation, Washington D.C., p 345-352

DeSalle R, Freedman T, Prager EM, Wilson AC (1987) Tempo and mode of sequence evolution in mitochondrial DNA of Hawaiin Drosophila. Journal of Molecular Evolution 26:157-164

Dixon MT, Hillis DM (1993) Ribosomal RNA secondary structure: compensatory mutations and implications for phylogenetic analysis. Mol Biol Evol 10:256-67

Edwards SV, Wilson AC (1990) Phylogenetically informative length polymorphisms and sequence variability in mitochondrial DNA of Australian songbirds (Pomatostomus). Genetics 126:695-711

Efron B (1982) The jackknife, the bootstrap adn otehr resampling plans.

Ehrlich M, Norris, K. F., Wang, R. Y. H., Kuo, K. C., Gehrke, C. W. (1986) DNA cytosine methylation and heat-induced deamination. Biosci. Rep. 6:387-393

Ehrlich M, Zhang XY, Asiedu CK, Khan R, Supakar PC (1990) Methylated DNA-binding protein from mammalian cells. In: Clawson GA, Willis DB, Weissbach A, Jones PA (eds) Nucleic acid methylation. Wiley-Liss, New York, p 351-366

Ekert R, Randall D (1983) Animal Physiology. 2nd ed. Freeman, New York

Eldredge N (1989) Macroevolutionary dynamics: species, niches & adaptive peaks. McGraw-Hill, New York

Ellsworth DL, Hewett-Emmett D, Li W-H (1993) Insulin-like growth factor II intron sequences support the hominoid rate-slowdown hypothesis. Mol. Phylogenet. Evol. 2:315-321.

Epstein CJ (1967) Non-randomness of amino-acid changes in the evolution of homologous proteins. Nature 215:355-359

Faith DPaPSC (1991) Could a cladogram this short have arisen by chance alone?: On permutation tests for cladistic structure. Cladistics 7:1-28

Fauron CMR, Wolstenholme DR (1980a) Extensive diversity among Drosophila species withrespect to nucleotide sequences within the adenine + thymine-rich region of mitochondrial DNA molecule. Nucleic Acids Res. 8: 2439-2452

Fauron CM-R, Wolstenholme DR (1980b) Intraspecific diversity of nucleotide sequences within the adenine+thymine-rich region of mitochondrial DNA molecules of Drosophila mauritiana, Drosophila melanogaster and Drosophila simulans. Nucleic Acids Res. 8: 5391-5410

Felsenstein J (1978) Cases in which parsimony and compatibility methods will be positively misleading. Systematic Zoology 27:401-410

Felsenstein J (1981) Evolutionary trees from DNA sequences: a maximum likelihood approach. Journal of Molecular Evolution 17:368-376

Felsenstein J (1985a) Confidence limits on phylogenies with a molecular clock. Systematic Zoology 34:152-161

Felsenstein J (1985b) Confidence limits on phylogenies: an approach using the bootstrap. Evolution 39:783-791

Felsenstein J (1985c) Phylogenies and the comparative method. Amer. Nat. 125: 1-15

Felsenstein J (1988a) Phylogenies and quantitative methods. Ann. Rev. Ecol. Syst. 19: 445-471

Felsenstein J (1988b) Phylogenies from molecular sequences: inference and reliability. Annu Rev Genet 22:521-65

Felsenstein J (1992) Estimating effective population size from samples of sequences: inefficiency of pairwise and segregating sites as compared to phylogenetic estimates. Genet Res 59:139-47

Felsenstein J (1993) PHYLIP 3.5 (phylogeny inference package). Department of Genetics, University of Washington, Seattle

Fitch WM, Margoliash E (1967) Construction of phylogenetic trees. Science 155:279-284

Fryer G (1998) A defense of arthropod polyphyly. In: Fortey RA, Thomas RH (eds) Arthropod relationships. Chapman & Hall, London, p 23-33

Gaut BS, Muse SV, Clark WD, Clegg MT (1992) Relative rates of nucleotide substitution at the rbcL locus of monocotyledonous plants. J. Mol. Evol. 35:292-303

Gillespie JH (1991) The causes of molecular evolution. Oxford University Press, Oxford

Goddard, J. M F, Fauron CM-R, Wolstenholme DR (1982) Nucleotide sequences within the A+T-rich region and the large-rRNA gene of mitochondrial DNA molecules of *Drosophila yakuba*. In: Slonimski P, Borst P, Attardi G (eds) *Mitochondrial genes*. Cold Spring Harbor Laboratory., p 99-103

Goddard JM, Wolstenholme DR (1980) Origin and direction of replication in mitochondrial DNA molecules from the genus Drosophila. Nucleic Acids Res. 8: 741-757

Gojobori T, Li WH, Graur D (1982) Patterns of nucleotide substitution in pseudogenes and functional genes. Journal of Molecular Evolution 18:360-369

Goldman N, Yang Z (1994) A codon-based model of nucleotide substitution for protein-coding DNA sequences. Mol. Biol. Evol. 11:725-736

Gouy M, Gautier C (1982) Codon usage in bacteria: correlation with gene expressivity. Nucleic Acids Res. 10: 7055-7064

Graessmann M, Graessmann A, Cadavid EO, Yokosawa J, Stocker AJ, Lara FJS (1992) Characterization of the elongation factor 1-α gene of *Rhynchosciara americana.* Nucleic Acids Research 20:3780

Grantham R (1974) Amino acid difference formula to help explain protein evolution. Science 185:862-864

Grantham R, Gautier C, Gouy M, Jacobzone M, Mercier R (1981) Codon catalog usage is a genome strategy modulated for gene expressivity. Nucleic Acids Res 9:r43-79.

Grantham R, Gautier C, Gouy M, Mercier R, Pave A (1980) Codon catalog usage and the genome hypothesis. Nucleic Acids Res 8:r49-79.

Gu X, Li W-H (1992) Higher rates of amino acid substitution in rodents than in humans. Mol. Phylogenet. Evol. 1:211-214

Gutell RR, Schnare MN, Gray MW (1990) A compilation of large subunit (23S-like) ribosomal RNA sequences presented in a secondary structure format. Nucleic Acids Res 18 Suppl:2319-30

Guttmann S, Boissonnas RA (1960) Synthèse de dix analogues de l'oxytocine et de la lysine-vasopressine contenant de la sérine, de l'histidine ou du tryptophane en position 2 ou 3. Helvetica Chimica Acta 43:200-216

Hartl DL, Moriyama EN, Sawyer SA (1994a) Selection intensity for codon bias. Genetics 1138:227-234.

Hartl DL, Moriyama EN, Sawyer SA (1994b) Selection intensity for codon bias. Genetics 138:227-34

Harvey PH, Keymer AE (1991) Comparing life histories using phylogenies. Philosophical Transactions of Royal Society of London, Series B 332:31-39

Harvey PH, Pagel MD (1991) The comparative method in evolutionary biology. Oxford University Press, Oxford.

Harvey PH, Purvis A (1991) Comparative methods for explaining adaptations. Nature 351:619-624

Hasegawa M, Kishino H, Yano TA (1985) Dating of the human-ape splitting by a molecular clock of mitochondrial DNA. Journal of Molecular Evolution 22:160-174

Hillis DM, Huelsenbeck JP (1992) Signal, noise, and reliability in molecular phylogenetic analyses. J Hered 83:189-95

Hochachka P (1991) Design of energy metabolism. In: Prosser. CL (ed) *Environmental and metabolic animal physiology*. Wiley-Liss, New York, p 353-436

Ikemura T (1981) Correlation between the abundance of *Escherichia coli* transfer RNAs and the occurrence of the respective codons in its protein genes: a proposal for a synonymous codon choice that is optimal for the E coli translational system. J Mol Biol 151:389-409.

Ikemura T (1982) Correlation between the abundance of yeast transfer RNAs and the occurrence of the respective codons in protein genes. J. Mol. Biol. 158:573-597

Ikemura T (1985) Codon usage and tRNA content in unicellular and multicellular organisms. Mol Biol Evol 2:13-34.

Ikemura T (1992) Correlation between codon usage and tRNA content in microorganisms. In: Hatfield DL, Lee B, Pirtle J (eds) Transfer RNA in protein synthesis. CRC Press, Boca Raton, Fla., p 87-111

Ikemura T, Ozeki H (1983) Codon usage and transfer RNA contents: organism-specific codon-choice patterns in reference to the isoacceptor contents. Cold Spring Harbor Symp. Quant. Biol. 47:1087

Ina Y (1995) New methods for estimating the numbers of synonymous and nonsynonymous substitutions. Journal of Molecular Evolution 40:190-226

Ingraham JL, MaalØe O, Neidhardt FC (1983) Growth of the bacterial cell. Sinauer, Sunderland, Mass.

Irwin DM, Kocher TD, Wilson AC (1991) Evolution of the cytochrome b gene of mammals. Journal of Molecular Evolution 32:128-144

James BD, Olsen GJ, Pace NR (1989) Phylogenetic comparative analysis fo RNA secondary structure. Meth. Enzmol. 180:227-239

Jaquenoud PA, Boissonnas RA (1959) Synthèse de la Phé^2-oxytocine. Helvetica Chimica Acta 42:788-793

Johnson NL, Kotz S, Kemp AW (1992) Univariate discrete distributions. John Wiley & Sons., New York

Jukes TH, Cantor CR (1969) Evolution of protein molecules. In: Munro HN (ed) Mammalian protein metabolism. Academic Press, New York, p 21-123

Kimura M (1980) A simple method for estimating evolutionary rates of base substitutions through comparative studies of nucleotide sequences. Journal of Molecular Evolution 16:111-120

Kimura M (1983) The neutral theory of molecular evolution. Cambridge University Press, Cambridge, England.

Kimura M, Ohta T (1972) On the stochastic model for estimation of mutational distance between homologous proteins. J. Mol. Evol. 2:87-90

Kishino H, Hasegawa M (1989) Evaluation of the maximum likelihood estimate of the evolutionary tree topologies from DNA sequence data, and the branching order in Hominoidea. Journal of Molecular Evolution 29:170-179

Kishino H, Miyata T, Hasegawa M (1990) Maximum likelihood inference of protein phylogeny and the origin of chloroplasts. Journal of Molecular Evolution 31:151-160

Kozak M (1983) Comparison of initiation of protein synthesis in procaryotes, eucaryotes, and organelles. Microbiol. Rev. 47:1-43

Krebs CJ (1999) Ecological methodology. Benjamin/Cummings., Menlo Park, CA

Kuhner MK, Felsenstein J (1994) A simulation comparison of phylogeny algorithms under equal and unequal evolutionary rates [published erratum appears in Mol Biol Evol 1995 May;12(3):525]. Mol Biol Evol 11:459-68

Kumar S, Tamura K, Nei M (1993) MEGA: Molecular Evolutionary Genetics Analysis. The Pennsylvania State

University, University Park, PA 16802.

Kurland CG (1987a) Strategies for efficiency and accuracy in gene expression1 The major codon preference: a growth optimization strategy. Trends Biochem Sci 12:126-128.

Kurland CG (1987b) Strategies for efficiency and accuracy in gene expression: 2 Growth optimized ribosomes. Trends Biochem Sci 12:169-171.

Kyte J, Doolittle RF (1982) A simple method for displaying the hydropathic character of a protein. Journal of Molecular Biology 157:105-132

Lake JA (1994) Reconstructing evolutionary trees from DNA and protein sequences: paralinear distances. Proceedings of National Academy of Sciences, USA 91:1455-1459

Latter BDH (1972) Selection in finite populations with multiple alleles. III. Geentic divergence with centripetal selection and mutation. Genetics 106:293-308

Lenstra JA, Van Vliet A, Carnberg AC, Van Hemert FJ, Möller W (1986) Genes coding for the elongation factor EF-1α in *Artemia.* European Journal of Biochemistry 155:475-483

Li W-H (1997) Molecular evolution. Sinauer, Sunderland, Massachusetts
Li WH (1993) Unbiased estimation of the rates of synonymous and nonsynonymous substitution. Journal of Molecular Evolution 36:96-99
Li WH, Graur. D (1991) Fundamentals of molecular evolution. Sinauer Associates Sunderland, Massachusetts
Li WH, Wu CI, Luo CC (1984) Nonrandomness of point mutation as reflected in nucleotide substitutions in pseudogenes and its evolutionary implications. J Mol Evol 21:58-71
Li WH, Wu CI, Luo CC (1985a) A new method for estimating synonymous and nonsynonymous rates of nucleotide substitution considering the relative likelihood of nucleotide and codon changes. Mol Biol Evol 2:150-74
Li W-H, Wu C-I, Luo C-C (1985b) A new method for estimating synonymous and nonsynonymous rates of nucleotide substitution considering the relative likelihood of nucleotide and codon changes. Mol Biol Evol 2:150-174
Liljenström H, vonHeijne G (1987) Translation rate modification by preferential codon usage: intragenic position effects. J. Theor. Biol. 124:43-55
Lockhart PJ, Steel MA, Hendy MD, Penny D (1994) Recovering evolutionary trees under a more realistic model of sequence evolution. Molecular Biology and Evolution 11:605-612
Martin AP (1995) Metabolic rate and directional nucleotide substitution in animal mitochondrial DNA. Mol. Biol. Evol. 12: 1124-1131
Mathieu O, R. Krauer, H. Hoppeler, P. Gehr, S. L. Lindstedt et al. (1981) Design of the mammalian respiratory system. VI. Scaling mitochondrial volume in skeletal muscle to body mass. Respir. Physiol. 44: 113-128
Miyata T, Hayashida H, Kuma K, Mitsuyasu K, Yasunaga T (1987) Male-driven molecular evolution: a model and nucleotide sequence analysis. Cold Spring Harb Symp Quant Biol 52:863-7
Miyata T, Miyazawa S, Yasunaga T (1979) Two types of amino acid substitutions in protein evolution. J-Mol-Evol 12:219-36
Miyata T, Yasunaga T (1980) Molecular evolution of mRNA: a method for estimating evolutionary rates of synonymous and amino acid substitutions from homologous nucleotide sequences and its application. J-Mol-Evol 16:23-36
Moriyama EN, Hartl DL (1993) Codon usage bias and base composition of nuclear genes in Drosophila. Genetics 134:847-58
Moriyama EN, Powell JR (1997) Synonymous substitution rates in Drosophila: mitochondrial versus nuclear genes. J-Mol-Evol 45:378-91
Morton BR, Clegg MT (1995) Neighboring base composition is strongly correlated with base substitution bias in a region of the chloroplast genome. Journal of Molecular Evolution 41:597-603
Muse SV, Gaut BS (1994) A likelihood approach for comparing synonymous and nonsynonymous nucleotide substitution rates, with application to the chloroplast genome. Molecular Biology and Evolution 11:715-724
Nagasenker PB (1984) On Bartlett's test fro homogeneity of variance. Biometrika 71:405-407
Nee S, Holmes EC, Rambaut A, Harvey PH (1996) Inferring population history from molecular phylogenies. In: Harvey PH, Brown AJL, Maynard Smith J, Nee S (eds) New uses for new phylogenies. Oxford University Press, Oxford, p 66-80
Nei M (1972) Genetic distance between populations. American Naturalist 106:283-292
Nei M (1987) Molecular Evolutionary Genetics. Columbia University Press, New York

Nei M (1991) Relative efficiencies of different tree-making methods for molecular data. In: Miyamoto MM, Cracraft J (eds) Phylogenetic analysis of DNA sequences. Oxford University Press, New York, p 90-128

Nei M, Gojobori T (1986) Simple methods for estimating the numbers of synonymous and nonsynonymous nucleotide substitutions. Molecular Biology and Evolution 3:418-426

Nei M, Tajima F, Tateno Y (1983) Accuracy of estimated phylogenetic trees from molecular data. II. Gene frequency data. J. Mol. Evol. 19:153-170

Noyer-Weidner M, Trautner TA (1993) Methylation of DNA in prokaryotes. In: Jost JP, Saluz HP (eds) DNA methylation: molecular biology and biological significance. Birkhäuser, Verlag, Basel, p 39-108

Ohta T, Kimura M (1973) A model o fmutation appropriate to estimate the number of electrophoretically detectable alleles in a finite population. Genet. Res. 22:201-204

Olson MS (1986) Bioenergetics and oxidative metabolism. In: Devlin TM (ed) Text book of biochemistry with clinical correlations. 2nd ed. John Wiley & Sons,, New York., p 212-260

Pace NR, Smith DK, Olsen GJ, James BD (1989) Phylogenetic comparative analysis and the secondary structure of ribonuclease P RNA--a review. Gene 82:65-75

Pedersen S (1984) *Escherichia coli* ribosomes translate in vivo with variable rate. EMBO J. 3:2895

Perler F, Efstratiadis A, Lomedico P, Gilbert W, Kolodner R, Dodgson J (1980) The evolution of genes: the chicken preproinsulin gene. Cell 20:555-66

Peters WS (1987) Counting for something: statistical principles and personalities. Springer-Verlag, New York

Regier JC, Shultz JW (1997) Molecular phylogeny of the major arthropod groups indicates polyphyly of crustaceans and a new hypothesis for the origin of hexapods. Molecular Biology and Evolution 14:902-913

Reynolds JB, Weir BS, Cockerham. CC (1983) Estimation of the coancestry coefficient: basis for a short-term genetic distance. Genetics 105:767-779

Rideout WMI, Coetzee GA, Olumi AF, Jones PA (1990) 5-Methylcytosine as an endogenous mutagen in the human LDL receptor and p53 genes. Science 249:1288-1290

Robinson M, R. Lilley, S. Little, J. S. Emtage, G. Yamamoto et al. (1984) Codon usage can effect efficiency of translation of genes in Escherichia coli. Nucleic Acids Res. 12: 6663-6671

Robinson M, Lilley R, Little S, others) as (1984) Codon usage can effect efficiency of translation of genes in Escherichia coli. Nucleic Acids Res 12:6663-6671

Rzhetsky A, Kumar S, Nei M (1995) Four-cluster analysis: a simple method to test phylogenetic hypotheses. Mol-Biol-Evol 12:163-7

Rzhetsky A, Nei M (1992a) A simple method for estimating and testing minimum-evolution trees. Molecular Biology and Evolution 9:945-967

Rzhetsky A, Nei M (1992b) Statistical properties of the ordinary least-squares, generalized least-squares, and minimum-evolution methods of phylogenetic inference. J-Mol-Evol 35:367-75

Rzhetsky A, Nei M (1993) Theoretical foundation of the minimum-evolution method of phylogenetic inference. Mol-Biol-Evol 10:1073-95

Rzhetsky A, Nei M (1994) METREE: a program package for inferring and testing minimum-evolution trees. Comput-Appl-Biosci 10:409-12

Saitou N, Nei M (1987) The neighbor-joining method: a new method for reconstructing phylogenetic trees. Molecular Biology and Evolution 4:406-425

Satta Y, Ishiwa H, Chigusa SI (1987) Analysis of nucleotide substitutions of mitochondrial DNAs in Drosophila melanogaster and its sibling species. Molecular Biology and Evolution 4:638-650

Schaaper RM, Danforth BN, Glickman BW (1986) Mechanisms of spontaneous mutagenesis: an analysis of the spectrum of spontaneous mutation in the E. coli lacI gene. J. Mol. Biol. 189:273-284

Seino S, Bell GI, Li W-H (1992) Sequences of primate insulin genes support the hypothesis of a slower rate of molecular evolution in humans and apes than in monkeys. Mol. Biol. Evol. 9:193-203

Sharp PM, Cowe E, Higgins DG, Shields DC, Wolfe KH, Wright F (1988) Codon usage patterns in Escherichia coli, Bacillus subtilis, Saccharomyces cerevisiae, Schizosaccharomyces pombe, Drosophila melanogaster and Homo sapiens; a review of the considerable within-species diversity. Nucleic Acids Res. 16:8207-11

Sharp PM, Devine KM (1989) Codon usage and gene expression level in Dictyostelium discoideum: highly expressed genes do "prefer" optimal codons. Nucleic Acids Res. 17:5029-5038.

Sharp PM, Li WH (1986) An evolutionary perspective on synonymous codon usage in unicellular organisms. J Mol Evol 24:28-38

Sharp PM, Li WH (1987) The codon adaptation index - a measure of directional synonymous codon usage bias, and its potential applications. Nucleic Acids Res 15:1281-1295.

Sharp PM, M. F. Tuohy, Mosurski KR (1986) Codon usage in yeast cluster analysis clearly differentiates highly and lowly expressed genes. Nucl. Acids Res. 14: 5125-5143

Singer CE, Ames BN (1970) Sunlight ultraviolet and bacterial DNA base ratios. Science 170:822-826

Smith RE (1956) Quantitative relations between liver mitochondria metabolism and total body weight in mammals. Ann. N. Y. Acad. Sci. 62: 403-422

Sneath PHA (1966) Relations between chemical structure and biological activity. Journal of Theoretical Biology 12:157-195

Sorensen MA, Kurland CG, Pedersen S (1989) Codon usage determines translation rate in Escherichia coli. J Mol Biol 207:365-377.

Srikantha T, Gutell RR, Morrow B, Soll DR (1994) Partial nucleotide sequence of a single ribosomal RNA coding region and secondary structure of the large subunit 25 s rRNA of Candida albicans. Curr Genet 26:321-8

Sved J, Bird A (1990) The expected equilibrium of the CpG dinucleotide in vertebrate genomes under a mutation model. Proc. Natl. Acad. Sci. USA 87:4692-4696

Swofford DL, Olsen GJ, Waddell PJ, Hillis DM (1996) Phylogenetic inference. In: Hillis DM, Moritz C, Mable BK (eds) Molecular Systematics. Sinaur, Sunderland, Mass., p 407-514

Tajima E, Nei M (1984) Estimation of evolutionary distance between nucleotide sequences. Molecular Biology and Evolution 1:269-285

Takezaki N, Nei M (1994) Inconsistency of the maximum parsimony method when the rate of nucleotide substitution is constant. J Mol Evol 39:210-218

Tamura K, Nei M (1993) Estimation of the number of nucleotide substitutions in the control region of mitochondrial DNA in humans and chimpanzees. Mol Biol Evol 10:512-526

Thomas WK, Beckenbach AT (1989) Variation in Salmonid mitochondrial DNA: evolutionary constraints and mechanisms of substitution. Journal of Molecular Evolution. 29:233-245

Thomas WK, Maa J, Wilson AC (1989) Shifting constraints on tRNA genes during mitochondrial DNA evolution in animals. New Biol 1:93-100

Thomas WK, Wilson AC (1991) Mode and tempo of molecular evolution in the nematode Caenorhabditis: cytochrome oxidase II and calmodulin sequences. Genetics 128:269-279

Thompson JD, Higgins DG, Gibson TJ (1994) CLUSTAL W: improving the sensitivity of progressive multiple sequence alignment through sequence weighting, positions-specific gap penalties and weight matrix choice. Nucleic Acids Research 22:4673-4680

Varenne S, Bug J, Lloubes R, Lazdunski C (1984) Translation is a non-uniform process: effect of tRNA availability on the rate of elongation of nascent polypeptide chains. J Biol Chem 180:549-576

Vogel F, Motulsky AG (1986) Human Genetics. Springer-Verlag, Berline and Heidelbert

Walldorf U, Hovemann BT (1990) *Apis mellifera* cytoplasmic elongation factor 1α (EF-1α) is closely related to *Drosophila melanogaster* EF-1α. FEBS 267:245-249

Weibel ER (1984) *The pathway for oxygen: structure and function in the mammalian respiratory system.* Harvard Univ. Press, Cambridge.

Wiebauer K, Neddermann P, Hughes M, Jiricny J (1993) The repair of 5-methylcytosine deamination damage. In: Jost JP, Saluz HP (eds) DNA methylation: molecular biology and biological significance. Birkhäuser Verlag, Basel, p 510-522

Williams DP, Rigier D, Akiyoshi D, Genbauffe F, Murphy JR (1988) Design, synthesis and expression of a human interleukin-2 gene incorporating the codon usage bias found in highly expressed Escherichia coli genes. Nucleic Acids Res. 16:10453-10467

Wu CI, Li WH (1985) Evidence for higher rates of nucleotide substitution in rodents than in man. Proc Natl Acad Sci U S A 82:1741-5

Xia X (1995) Body temperature, rate of biosynthesis and evolution of genome size. Mol Biol Evol 12:834-842

Xia X (1996) Maximizing transcription efficiency causes codon usage bias. Genetics 144:1309-1320

Xia X (1998a) How optimized is the translational machinery in *E. coli*, *S. typhimurium*, and *S. cerevisiae*? Genetics 149:37-44

Xia X (1998b) The rate heterogeneity of nonsynonymous substitutions in mammalian mitochondrial genes. Molecular Biology and Evolution 15:336-344

Xia X, Boonstra R (1992) Measuring temporal variation in population density: a critique. American Naturalist 140:883-892

Xia X, Hafner MS, Sudman PD (1996) On transition bias in mitochondrial genes of pocket gophers. Journal of Molecular Evolution 43:32-40

Xia X, Li W-H (1998) What amino acid properties affect protein evolution? Journal of Molecular Evolution. 47:557-564

Yang Z (1994) Estimating the pattern of nucleotide substitution. J Mol Evol 39:105-111

Yang Z (1996a) Among-site rate variation and its impact on phylogenetic analysis. TREE 11:367-372

Yang Z (1996b) Maximum-likelihood models for combined analyses of multiple sequence data. J. Mol. Evol. 42:587-596

Yang Z (1997a) How often do wrong models produce better phylogenies? Mol Biol Evol 14:105-108

Yang Z (1997b) PAML: a program package for phylogenetic analysis by maximum likelihood. CABIOS 13:555-556

Yang Z, Kumar S, Nei M (1995) A new method of inference of ancestral nucleotide and amino acid sequences. Genetics 141:1641-1650

Yang Z, Nielsen R, Hasegawa M (1998) Models of amino acid substitution and applications to mitochondrial protein evolution. Molecular Biology and Evolution 15:1600-1611

Zar JH (1996) Biostatistical analysis. Prentice-Hall, Englewood Cliffs, N.J.
Zharkikh A (1994) Estimation of evolutionary distances between nucleotide sequences. Journal of Molecular Evolution 39:315-329
Zischler H, Geisert H, Haeseler Av, Pääbo S (1995) A nuclear 'fossil' of the mitochondrial D-loop and the origin of modern humans. Nature **378**:489-492
Zuckerkandl E, Pauling L (1965) Evolutionary divergence and convergence in proteins. In: Bryson V, Vogel HJ (eds) Evolving genes and proteins. Academic Press, New York, p 97-166

Index

MIX
Papier aus verantwortungsvollen Quellen
Paper from responsible sources
FSC® C105338

If you have any concerns about our products,
you can contact us on
ProductSafety@springernature.com

In case Publisher is established outside the EU,
the EU authorized representative is:
Springer Nature Customer Service Center GmbH
Europaplatz 3, 69115 Heidelberg, Germany

Printed by Libri Plureos GmbH
in Hamburg, Germany